AF594270

A Strategy for Occupational Exposure Assessment

A Strategy for Occupational Exposure Assessment

Edited by
NEIL C. HAWKINS
SAMUEL K. NORWOOD
JAMES C. ROCK

American Industrial Hygiene Association
Akron, Ohio

ISBN 0-932627-46-3

Published by American Industrial Hygiene Association
Mary Turbak, Editor, Publications
345 White Pond Drive, P.O. Box 8390
Akron, OH 44320

AIHA Exposure Assessment Strategies Committee 1986–1989

1988–1989 Members (voting to approve consensus document)

Lawrence R. Birkner
Christopher J. Cole
Joe Damiano
Nurtan Esmen
George Flores
Doan J. Hansen
Neil C. Hawkins
Richard S. Holmes
Dennis P. Johnson
Bruce Larson
Roger Lewis
Peter L. Lubs
Frederick J. Molohon
Thomas J. Nelson
Samuel K. Norwood
Frank M. Parker III
Colonel James C. Rock
Bernard E. Saltzman
A. Robert Schnatter
Paul Schubert
Ernest L. Timlin, Jr.
R. Mike Tuggle
James L. Unmack
Constance Wrench
Richard E. Zimmerman

1986–1989 Consultants and Other Committee Members

D.L. Bosatra
Gerald L. Cooper
John S. Evans
Janice D. Florin
R.L. Harris
N.A. Leidel
J.R. Lynch
C.R. "Gus" Manning
S. Roach
Alfredo Salazar
U. Ulfvarson

Contents

Preface

This document represents a consensus developed over a 3-year period by many professionals from the American Industrial Hygiene Association (AIHA). The original advocacy and call to form an AIHA Exposure Assessment Strategies Committee came from John Henshaw. The committee organized with 10 members and elected officers at the 1986 American Industrial Hygiene Conference (AIHC). Its first effort involved cosponsoring, along with the American Petroleum Institute and the Chemical Manufacturers' Association, an international Workshop on Strategies for Measuring Exposure. At the 1987 AIHC, the committee sponsored the best-attended Friday morning session ever. The roundtable, "Exposure Assessment Strategies," generated 10 formal presentations and much meaningful discussion. The 1989 AIHC featured a second session—11 very strong papers, each supported by a six-page extended abstract, available to conference attendees. These events helped ensure that the committee had not overlooked any concepts useful to its strategy document and laid the foundation for subsequent deliberations.

The committee has met for 2-day writing and editing workshops three different times. After each meeting the revised draft was mailed to all members for comment and appropriate revisions. The committee developed and presented a professional development course based on a draft of this manual. The final editing also incorporated ideas received from extensive comments from other AIHA committees and from board members.

This document is intended for industrial hygienists. It is a handbook of practice for workplace exposure assessments conducted by industrial hygienists and is intended to provide a common language as well as a basis for professional practice. As such, it is also useful to other members of the occupational health team who need to understand and interpret workplace exposure assessment data.

While seeking the necessary consensus among committee members, one great challenge has been terminology. The committee discovered it could not reach consensus unless it first agreed on definitions of terms. The reader is advised to read the glossary of terms to facilitate proper understanding of the concepts presented in the text.

The Exposure Assessment Strategies Committee believes this manual to be the first attempt to reach consensus on the most valued tools of industrial hygiene—the occupational exposure limit, the concept of overexposure, the analytical efficiency of homogeneous exposure groups, and, most important, the professional judgment of the industrial hygienist who uses these tools to ensure a safe and healthful workplace. Each of these key concepts is addressed in more detail in Chapter 1.

The committee offers this strategy document as a living, evolving document designed to be improved in subsequent editions. The committee's future endeavors will emphasize new editions of the basic document as well as detailed supplemental documents for the intricate aspects of this overall strategy.

The committee invites individuals who wish to provide their unique insight in this field to address comments to the chair of the AIHA Exposure Assessment Strategies Committee. Better yet, such individuals are invited and encouraged to join the committee to help define the elements of practical and useful exposure assessment strategies.

NEIL C. HAWKINS
SAMUEL K. NORWOOD
JAMES C. ROCK

Acknowledgment

The committee thanks Connie Haubenstricker for her outstanding efforts in compiling the early drafts of this document (all text and figures), Irma Ledesma for later drafts of text, and Tom Holder for some of the final figures. This document would have been impossible without them!

1
Introduction

The Purpose of the Occupational Exposure Assessment Strategy

Industrial hygienists recognize, evaluate, and control potential workplace hazards so that all employees are assured a safe and healthful occupational environment throughout their working lifetimes. As today's workplaces grow more complex, new challenges abound for industrial hygienists. Evaluation of these complex situations requires a sound, logical workplace exposure assessment strategy to focus industrial hygiene and other occupational health resources on those work situations with the greatest potential for adverse health effects. The industrial hygienist must assess workers' exposures to chemical, physical, and biological agents in light of the potential health effects of these agents in order to assess the risks associated with working in that environment. The purpose of devising a sound exposure assessment strategy is to accomplish at least four goals:

1. To assess potential health risks faced by all workers, to differentiate between acceptable and unacceptable exposures, and to control unacceptable exposures
2. To establish and document a historical record of exposure levels for all workers and to communicate exposure monitoring results to each worker
3. To ensure and demonstrate compliance with governmental and other exposure guidelines
4. To accomplish the above three goals with efficient and effective allocation of time and resources

The Basic Problem

The basic problem in occupational exposure assessment is to recognize all exposures, to evaluate each as acceptable or unacceptable, and to control all unacceptable exposures. Defining homogeneous exposure groups (HEGs)—a complex task—helps to accomplish each of these objectives. Throughout the process of recognizing, evaluating, and controlling exposures, the use of professional judgment to plan monitoring campaigns, to interpret data, and to recommend changes is essential.

Recognition of Exposures

Recognizing exposures involves considering all factors shown in Figure 1.1. Workers may be exposed to environmental agents (chemical, physical, and biological) during the performance of tasks directly involving these agents, by incidental contact from background contamination in the workplace, or by the effects of nearby tasks performed by others. Although the highest exposures usually result from a worker's own tasks, significant exposures do occur from the other two sources. A credible occupational exposure assessment strategy considers all sources of exposure.

Evaluation of the Significance of Exposures

To evaluate the significance of exposures, both the time history of each exposure and the occupational exposure limit for the environmental agent of concern must be understood. Modern workplaces are complex—concentrations of environmental agents vary over both time and space. Workers move through these variable concentrations on variable paths and perform tasks that affect the concentrations. Most deterministic models of employee exposure are too complex to be of practical use. Therefore, most industrial hygiene decisions are based on a few representative exposure measurements. The acceptability of a workplace is judged by comparing monitoring results with occupational exposure limits (OELs) and by estimating by some means the proportion of overexposures occurring in that

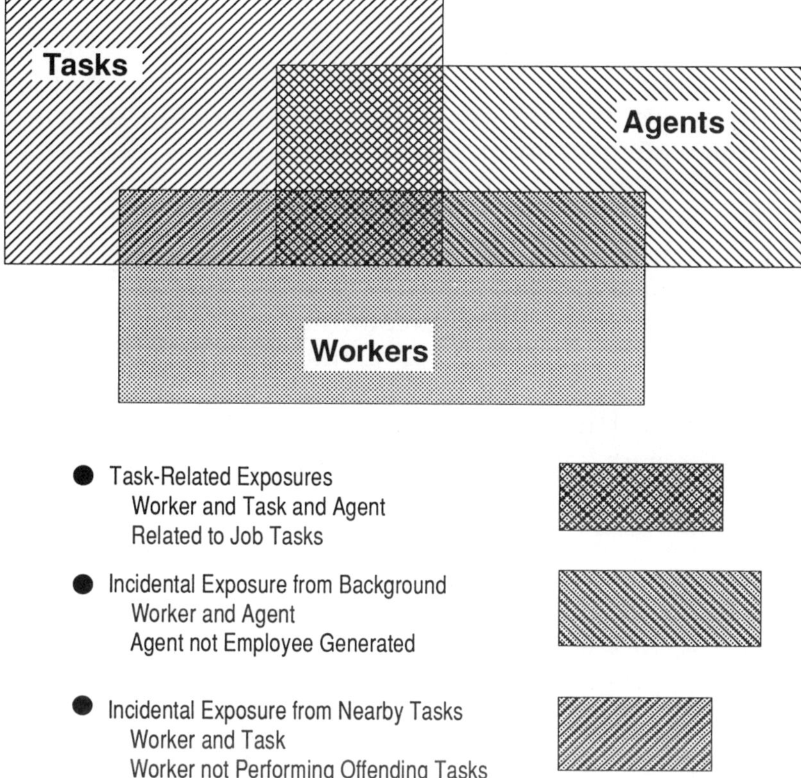

Figure 1.1. Workplace exposure factors

workplace. Though this has been a rather subjective process, the consensus is that now is the time for industrial hygienists to standardize their approach. To do so, they must be rigorous in their definitions and in the use of the terms *occupational exposure limits* and *overexposure.*

Occupational Exposure Limits Occupational exposure limits must be expressed in terms that are consistent with data generated from workplace exposure monitoring. All monitoring results are averages. Airborne agent exposures are measured as an average concentration (chemical agent) or average intensity (physical agent) observed during the sampling time. Surface exposures are measured as average concentration over the surface area sampled. Thus, an OEL is a pair of numbers: either a concentration and an averaging time or a quantity and a surface area.

Setting proper values for OELs is both complex and controversial. *Occupational exposure limit* is defined as a guideline for good practice. It is not a fine line separating absolute safety from impending doom. Rather, the OEL represents an exposure to which most workers can be exposed regularly, day after day, without adverse health effects. Some environmental agents may have more than one OEL assigned. To protect against inhalation hazards, the OELs for some agents specify one concentration for short exposures (acute health effect from high concentrations) and a lower concentration for long exposures. A surface concentration is sometimes specified for agents that can be absorbed through the skin. Acceptable exposure patterns for agents with multiple OELs require all OELs to be met simultaneously.

An Overexposure An overexposure is observed when monitoring demonstrates an average concentration greater than an OEL *and* the sample was averaged as stated in the OEL (proper duration or proper surface area). An overexposure may also be estimated by statistical tools that predict an unacceptable level of exposure for the stated time period (i.e., several 15-min samples used to estimate an 8-hr average exposure). Only when the monitoring results represent averages over the time period (or the area) specified in the OEL can the comparison be made. Averages taken over differing times or surface areas must not be compared with one another or with an OEL. Each overexposure observed should be thoroughly investigated to identify and correct its cause.

Control of Unacceptable Exposures

To control unacceptable exposures, obtaining considerable detail about the workplace is necessary. The full process of exposure must be evaluated, including the source, the pathway, and the worker. Control measures can be applied at any step. Process changes can reduce source emission strengths, and ventilation system adjustments can intercept contaminants and eliminate the pathway. Finally, personal protective equipment, properly used by a worker, provides an additional protection factor—but is appropriate only when all other measures are inadequate.

Homogeneous Exposure Groups

A homogeneous exposure group (HEG) is the central philosophical construct for detailed workplace exposure assessments. In its purest conceptual form, an HEG is a group of workers with identical probabilities of exposure to a single environmental agent. The group is homogeneous in the sense that the probability distribution of exposures is the same for all members of the group—all members of the group need not have identical exposures on any single day. Because of the statistical homogeneity, a small number of randomly selected samples can be used to define exposure distributions and trends within the HEG. Thus, the HEG forms the basis for quantitative industrial hygiene. The HEGs are set up during the identification step, they are characterized during the evaluation step, and they receive needed attention during the control step.

Professional Judgment

Professional judgment is a term that defies precise definition but is nonetheless required for successful occupational exposure assessment. It is essential in planning monitoring campaigns (recognizing the hazard), in interpreting resultant exposure data (evaluating the hazard), and in recommending needed process changes (controlling the hazard). Professional judgment is essential to identifying HEGs. Exposure data from each HEG must provide unbiased estimates both of the distributions of exposure intensities and of the exposure trends over time. Professional judgment is also required while evaluating exposure distributions.

The flexibility of the exposure assessment strategy not only permits but forces subjective decisions based on professional judgment. Throughout this text both qualitative and quantitative comments have been provided to describe the professional judgment underlying proper application of the essential elements of an exposure assessment strategy. These descriptions are intended to help all industrial hygienists to improve their own professional judgment.

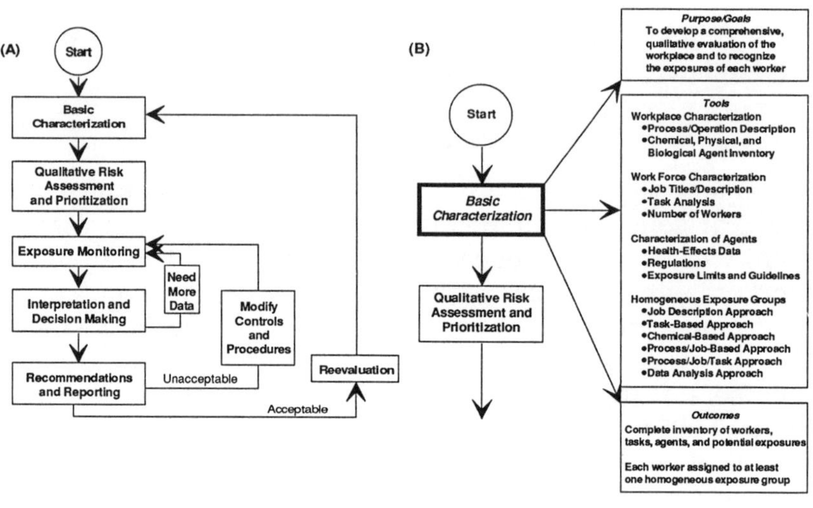

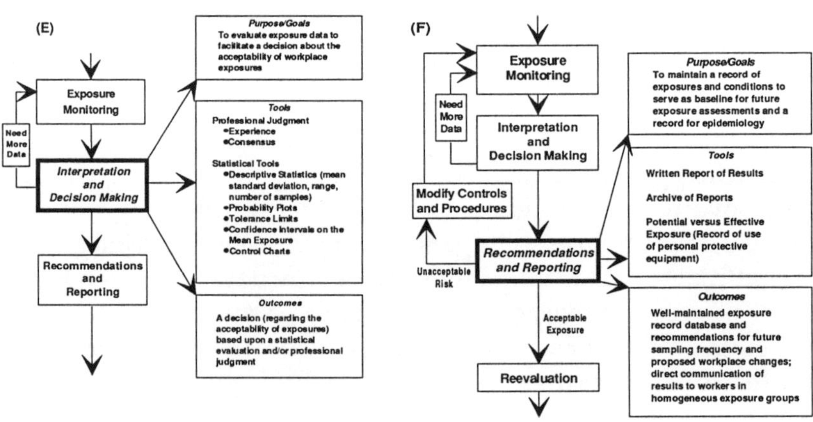

Overview of the Exposure Assessment Strategy

Figure 1.2A presents an overall flow diagram of the occupational exposure assessment strategy. This simple (but powerful) strategy is suitable for evaluating many occupational environments, work forces, and agents. Figure 1.2B–G presents in more

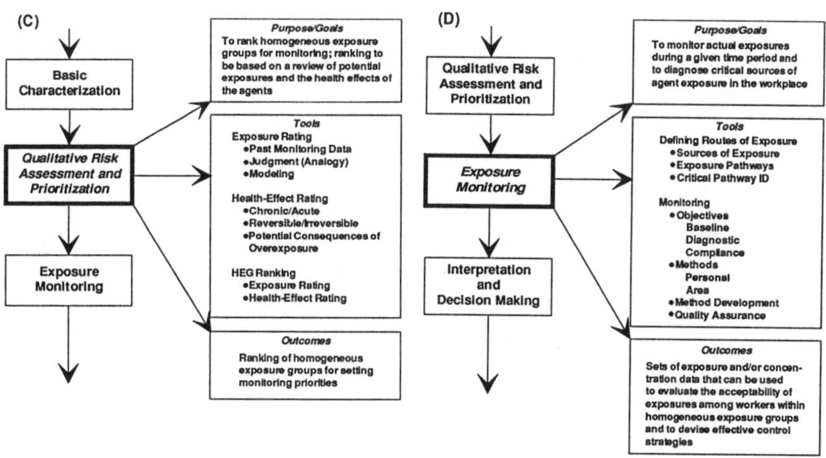

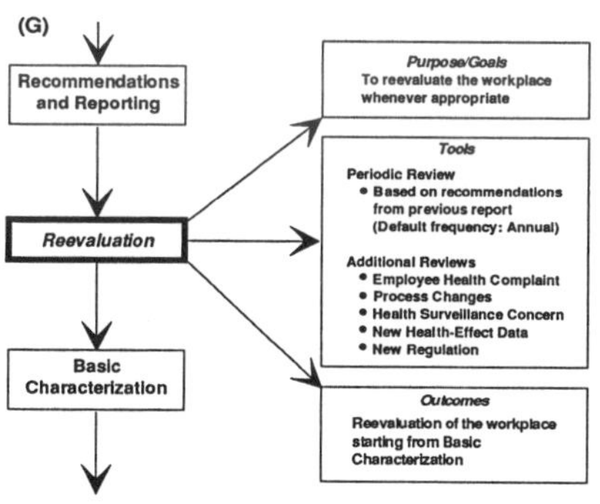

Figure 1.2 The occupational exposure assessment strategy and expanded views of each block (strategy component). A = Overall flow diagram of the strategy, B = Basic characterization component, C = Qualitative risk assessment and prioritization component, D = Monitoring component, E = Interpretation and decision-making component, F = Recommendations and reporting component, and G = Reevaluation component.

detail the goals, tools, and outcomes of each block (strategy component) of the overall strategy.

Careful study and comparison of the overall strategy and the details of any block of the strategy will provide a convenient overview or review of the central themes of the strategy.

SUMMARY

The general flow of this exposure assessment strategy presents the logical steps an experienced industrial hygienist follows. The outcomes of performing each of these steps provide answers to the following questions.

- What are the potential exposures?
- Who is exposed and at what level? What are the effective exposures considering the protection factors of personal protective equipment, as used?
- How toxic are the agents?
- Which groups require priority in a monitoring campaign?
- Do sampling results indicate that the risk is acceptably low? If not, controls should be implemented.
- Once results have been reported, when does one need to reevaluate?

This document is simply a written version of what many industrial hygienists already do in practice.

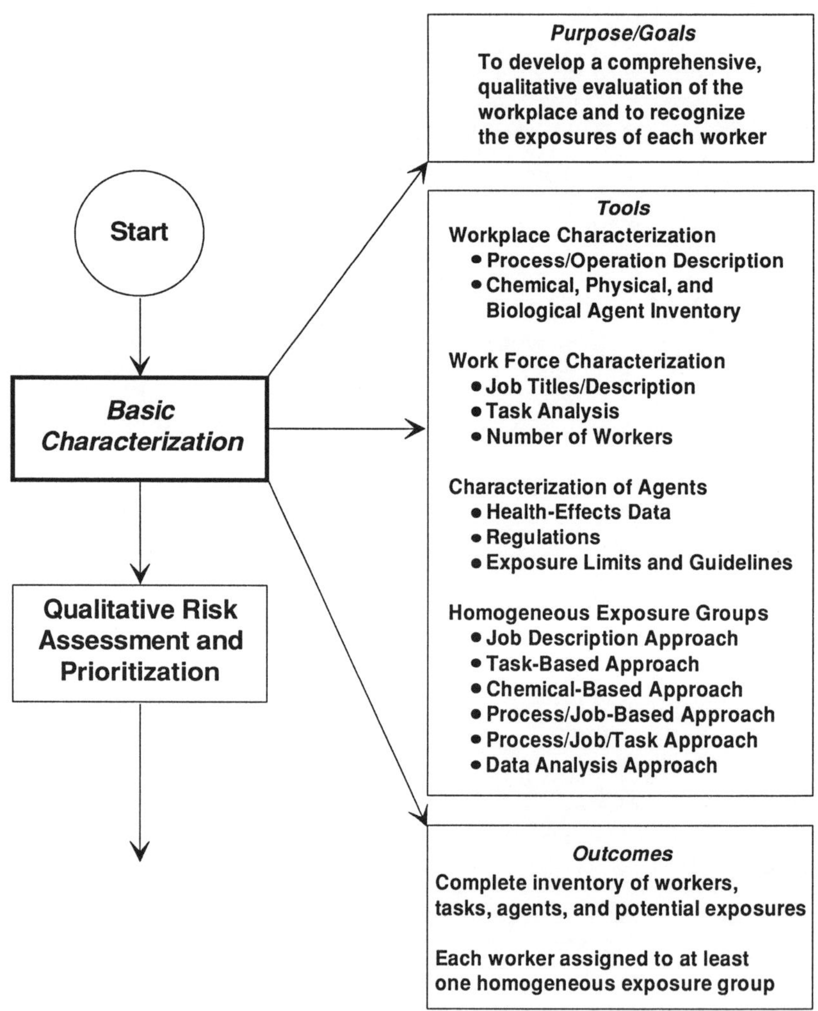

Figure 2.1. Basic characterization component of the exposure assessment strategy

2
Basic Characterization

Purpose

The first step in an occupational exposure assessment is to characterize the workplace. This basic characterization should identify potential exposures for every employee assigned to the workplace, should identify the occupational exposure limit (OEL) appropriate for each exposure, and should define the homogeneous exposure groups (HEGs).

Overview of Basic Characterization Tools

The identification of exposures involves detective work to describe the work force fully; the processes, operations and tasks; and a complete inventory of chemical, biological, and physical agents with appropriate OELs. Workplace and work force characterization are used as the basis for assigning each employee to an HEG. The notion of an HEG is useful in that once the workers are properly classified, any monitoring result for a worker within a group is presumed to provide information representative of the exposures of other workers from that group.

Figure 2.1 illustrates aspects of conducting a basic characterization.

The industrial hygienist must have a sufficient understanding of the work environment to make an evaluation of potential health risk. At a minimum, the following questions must be answered.

1. What are the chemical, biological, and physical agents in the work environment?

2. What are the health effects associated with excessive exposure to the agents?
3. What are the OELs for each agent?
4. How are workers and groupings of workers in the work environment involved with each of the agents?
5. What are the key operations and activities that pose the greatest potential for exposure to the agents?

Thorough characterizations of the workplace, the work force, and the environmental agents are needed before the industrial hygienist can develop a program to address the agents in some reasonable order of priority. Priority setting is important for effective utilization of the limited resources available to conduct monitoring programs. In addition, this basic characterization will provide historical records of the process conditions and job activities that may be useful for long-range health surveillance.

The basic characterization has four major components:

1. Workplace characterization
2. Work force characterization
3. Agent characterization
4. Determination of HEGs

This strategy can be used for a wide variety of workplace situations. A hypothetical example (sodium chloride production plant) is presented in Appendix I to illustrate components of basic characterization as well as the complete exposure assessment strategy.

Workplace Characterization

Most workplaces can be adequately described by preparing a schematic diagram and/or a written description of the process or operation. The purpose of these outlines is to highlight key unit operations, activities, and/or areas with exposure potential and to provide details of the production activities and process chemistry. In some cases, the written description provides a mechanism for defining a work environment that is not in a fixed location (transportation, maintenance, janitorial work, etc.) or that is unusual in nature (exposures of artists in a workshop).

Schematic Diagram A schematic diagram is a useful tool in evaluating any workplace's potential exposures. The use of schematic diagrams is equally critical in describing chemical industry exposures, manufacturing and fabrication plant exposures, service industry exposures, and even indoor air quality exposures. The goal is to identify the sources of potential exposure, locations of chemicals and tasks, and ventilation and other control methods. Chemical and physical process information should be documented as needed.

From an industrial hygiene standpoint, the key elements of a process flow diagram are the sources of potential exposure. A source of potential exposure is a piece of equipment and/or worker task that potentially brings a worker into contact with the environmental agent. For this reason, small equipment and associated tasks (pumps, in-line filters, sampling points, etc.) may need to be included in the flow diagram. Likewise, a material entering or leaving the process should be classified by the type of transfer (pipeline, tank truck, drumming, bagging, etc.) involved since this highlights possible handling problems that could lead to exposure. The transfer method and disposition of all by-product streams leaving the process (e.g., noncondensable vapors to vent, waste water to treatment system via trench, etc.) should be shown to indicate possible sources of exposure and to provide a resource for other environmental studies. With the proper review, a well-prepared flow diagram will spotlight the main areas of concern and will also minimize the possibility that potential hazards will be overlooked. A flow chart/diagram allows the industrial hygienist to summarize the facility activities, as they relate to potential exposures, with a minimum of effort.

Workplace Description A workplace description should be written to explain what activities are taking place in the work area. Sometimes complex chemical and physical processes must be described, but in other situations only basic processes, such as evaporation of a solvent into a room, need to be noted. Local exhaust ventilation and general dilution ventilation should be described. The workplace description should emphasize where potential exposures may occur; it may be a very simple description (for example, the workplace exposure situation involves only offgassing of formaldehyde from particleboard into a room with recirculated dilution air). On the other hand, the workplace description may be more complex such as in a batch chemical processing plant or a semiconductor manufacturing operation. Concentrations of environmental agents leaving the workplace in exhaust ventilation should not be overlooked. One must verify that such streams do not reenter through fresh air intakes.

While providing useful information, the flow diagram cannot give a complete picture of what is occurring in the process; personal knowledge of chemistry and physics and experience are necessary to describe the various reactions that may be taking place. The written process description should minimize any confusion by describing how the raw materials are being changed and what products, intermediates, or by-products are created at various stages of the operation. A thorough understanding of the process chemistry is essential for the industrial hygienist to make a complete assessment of the potential health risks in any workplace involving chemical processes.

The process description should include important details such as component stream concentrations, operating temperatures, and pressures. These data can be used to evaluate which chemical components pose the greatest potential for exposure. Other factors that affect the potential for exposure (ventilation systems, open-top tanks, open sumps or trenches, use of protective equipment) should be included in the description. If respirators are used during high exposure tasks, the industrial hygienist should clearly differentiate potential exposures (breathing zone concentrations) from effective exposures (reduced by the respirator protection factor).

The description can also be used to provide information on auxiliary operations (material handling, refrigeration units, maintenance, etc.) that may involve exposure to nonprocess agents. For later historical analyses, an indication of the current operating level (tonnage, percent of capacity, etc.) could provide some perspective on overall exposure potential, particularly if the operating data are compared with related monitoring results. Although brief, the process description complements the process flow diagram and provides many of the essential details to support the basic workplace characterization by the industrial hygienist.

Work Force Characterization

To understand how plant personnel interface with the process and its agents, the industrial hygienist must gain a detailed knowledge of the operating procedures and work practices used in the plant. Every effort should be made to use existing information about the work force (e.g., plant roster, job descriptions, and worker and management interviews) to help subdivide the people into groups of similar exposure profiles. In this way, the industrial hygienist can quickly define the groups with higher exposure potential that may need to be monitored more intensively than groups with lower potential.

Job Classifications / Assignments / Number of Workers Assigned Many organizations have personnel record systems that provide a roster of individuals, along with a summary of job titles, for specific operating units, plants, and/or departments. The roster should be reviewed to identify groups having the same or very similar work duties; in most cases, these groupings may include several distinct job classifications.

Even if a grouping has only one job classification, work duties need to be reviewed carefully by the industrial hygienist to make sure that each group is adequately characterized for unique exposure potential. In some cases, job classifications must be subdivided into assignments so that significant differences in job activities or tasks can be easily related to the specific individuals. If the personnel record system cannot be used to maintain the assignment information, the industrial hygienist should develop a permanent file defining the individual, the job classification, and the specific assignment/task that distinguishes this individual from others in the job category for each basic characterization. If the industrial hygienist deviates from the personnel system for describing work groups or tracking work history, the industrial hygienist's responsibility for maintaining accurate records becomes more important should epidemiologic studies ever be performed. Incongruence of industrial hygiene classifications and worker tracking systems could make future epidemiologic studies impossible.

Job Descriptions/Task Analysis For each of the job classification/assignment groups, a brief outline of work duties and tasks is needed. For groups that have only a negligible exposure potential (supervisory, clerical, computer support, control room operators, etc.), the work duties may be adequately covered by defining the percent of time spent in the general process areas. However, for groups intimately involved with the operation of the plant, more details are needed about the specific process areas, the work activities conducted in each area, and the frequency of those activities (often called *task analysis*). By observation and worker interview, the specific unit operations and tasks that lead to potential worker exposure should be identified, as well as the frequency and duration of these tasks (e.g., once per hr for 10 min, once per shift for 30 min, or twice per year for 4 hr).

The detailed work area/activity description can be prepared in the form of a time distribution, showing each major work area (reactor train, bagging station, analytical lab, etc.) and the time spent on each work activity (task) within that area. The main activities (process sampling, product bagging, titration analysis, etc.) should be included, particularly if a significant potential for exposure exists. Activities that are done infrequently should be described with both frequency and duration estimates. For example, a task may be performed once per month and take 4 hr to complete.

Demographic and Life Style Data In addition to the job descriptions, other data related to the individual may be needed to evaluate total exposure potential. Work history data, including all job classifications and assignments, are essential for long-term health assessments; the personnel department generally maintains this type of information. The medical department may have information related to life style factors such as smoking habits, recreational activities (hunting, motorcycling, etc.), and outside occupations (farming, carpentry, etc.) that may contribute to the total exposure burden (noise, chemicals, etc.).

The industrial hygienist must be familiar with all of the resources that can be utilized within an organization to make completed evaluations of exposure potential. While not commonly used in the day-to-day development of monitoring strategies, demographic and life style data may be important factors in an epidemiologic study or in a specific investigation involving adverse health effects. The committee recommends such data be collected only if confidentiality can be maintained or where workers agree as to present and future uses of the data. This may be difficult for small firms that have no medical group and only minimal or part-time industrial hygiene support.

Characterization of Environmental Agents

The workplace and work force characterizations involve data gathering so that the industrial hygienist can apply judgment in deciding which job classifications have the highest potential for exposure. However, the ultimate goal is to estimate the risk of an unacceptable workplace based on the potential exposure and the health effects of the agents encountered in the workplace. By assessing the potential for exposure and the health effects, the industrial hygienist can determine which agents need to be monitored and which groups of workers should be evaluated with the highest priority.

To evaluate the potential for exposure, the workplace and work force data must first be organized so that the potential exposure can be estimated for each job classification/agent combination; some data on physical properties and amounts handled may also be required for each agent to permit the modeling of worst-case exposure potential. In addition, to complete a qualitative risk assessment (potential for an unacceptable workplace), health-effects data, guidelines for acceptable levels of exposure, and past exposure data must be gathered for each agent. These will be used in the procedure for setting priorities (described in the next chapter).

Chemical, Physical, and Biological Agents Inventory The process and job descriptions and other knowledge of the workplace can be used to prepare a relatively complete list (inventory) of the agents that may be encountered in the work environment. A separate list should be made for each HEG that was identified as having a unique exposure potential. Typically, these inventories will break down first by process area, second by job description/category, and third by agents faced by those workers in that specific process. If this list is to be used only for priority setting, the industrial hygienist can use judgment to exclude some agents that meet the following criteria: (1) only

TABLE 2.1
Categories of Environmental Agents in Chemical Processing Operations

Category	Description
Raw materials	Primary reactants that often comprise the largest quantities of materials handled
Catalysts	Materials that initiate the main reactions and usually remain unchanged; quantities used are usually small (<10%) in comparison to the raw materials.
Auxiliary materials	Nonreactants (solvents, slurry media, filter aids, etc.) that support the main reaction; generally handled in quantities similar to the raw materials
Products	Desired compounds formed by the main reactions, in quantities similar to the raw materials, which usually have to be separated from the auxiliary materials
By-products	Secondary materials formed by the main reactions, usually in trace quantities, which usually have to be separated from the products as waste streams
Additives	Supplementary materials (inhibitors, surfactants, pigments, etc.) added to the final product to enhance the performance; usually in small quantities (<1%)
Lab chemicals	Materials used for the preparation of samples or performance of wet chemistry and instrumental analyses; usually handled in gram or milliliter quantities
Biological agents	Organisms and biologically active materials produced by living organisms. Examples include spores, casts, endotoxins, etiological agents, etc.
Construction materials	Refractories, solvents, insulations, welding fumes, lubricants, adhesives, etc. to which maintenance workers are routinely exposed

small amounts available for contact (e.g., gram quantities or quantities small in comparison with a threshold dose); (2) infrequent use (e.g., less than once per month); or (3) low agent toxicity.

The inventory should be sorted in a manner that gives an indication of how each agent is used in the workplace. One classification of the process materials that may provide some perspective on the amounts handled includes Table 2.1. This system is commonly used to sort chemical process agents, but any system that meets the needs of the operation can be employed as long as the sorting is done consistently.

Quantity/Physical Properties While categories in the chemical, physical, and biological agents inventory suggest the relative amounts of material that are handled, the industrial hygienist must also have an understanding of the actual quantities that are used in the process. Exact amounts are not required, since only a general range (grams, pounds, tons, etc.) is needed for the assessment of potential for exposure. Bulk quantity information is often available from warehouse records.

Physical property data (boiling points, vapor, pressure, particle size distribution, etc.) for each of the agents, under condition of use, are also needed to assess the potential for exposure. For liquid streams, Raoult's law for ideal solutions can be used to estimate the saturated vapor concentrations for the various components; the saturated vapor concentrations, in turn, indicate which materials will be the main components of any airborne vapors. The vapor pressure and temperature will also assist in determining whether the contaminant will exist as a vapor, an aerosol, or both. Both physical forms should be considered in determining inhalation potential. For solid materials, knowledge of the particle size distribution generated during worker contact will help the industrial hygienist to evaluate the inhalation potential of dust exposures.

Health-Effects Data The chemical, physical, and biological agents inventory can be used by the industrial hygienist as a checklist to make sure that health-effects data are available for each agent. At a minimum, industrial hygienists should obtain material safety data sheets for agents. The lack of data may prompt some toxicologic testing depending on the importance of the agent in the workplace. The industrial hygienist may request that the manufacturer of the substance perform toxicologic studies or may initiate internal testing. An interpretation of the available data should be prepared for presentation to workers and management.

To be helpful as part of the qualitative risk assessment, the health-effects data must be sufficient to estimate acceptable levels of exposure. In many cases, this step is simplified by existing guidelines established by regulatory or consensus organizations. Examples include the Occupational Safety and Health Administration's permissible exposure limits (PELs), the American Conference of Governmental Industrial Hygienists' threshold limit values (TLVs®), and the American Industrial Hygiene Association's workplace environmental exposure level (WEEL) and hygienic guides. In this document, the generic term *occupational exposure limit* was used to represent the selected guideline, whether one of the above or an internal company guideline based on plant level experience.

Even if an exposure guideline is available, the industrial hygienist must still understand the reasons why the guideline was set at a particular level. For the qualitative risk assessment (potential for health effects), more emphasis should be placed on a guideline that was based on an insidious and irreversible health effect (carcinogenesis, internal organ injury, birth defects, etc.) than one based on nuisance or irritant properties.

The review of the toxicology data should lead to conclusions for several questions.

1. Does the agent have a time dependence (dose-rate effect)? Does chronic low level exposure result in the same effects as brief high level exposures when both result in the same cumulative dose?
2. Are the health effects reversible or irreversible? Do the health effects include reproductive, teratogenic, or carcinogenic effects; immediate death; or disability?
3. What is the relevant OEL in light of the potential health effects and rate dependence? What concentration is appropriate? What sample averaging time is appropriate for the OEL?

4. Does the agent have warning properties (odor or irritation) at levels less than the exposure guideline or only at levels greater than the guideline? Does the agent have warning properties at levels less than or greater than concentrations that are immediately dangerous to life or health?
5. What safety factor was used in deriving the OEL and were the health-effect results based on animal or human studies? Were the health-effect studies acute or chronic and how do these compare to actual worker exposures? How strong are the health data that support the exposure guideline?

The answers to these questions are critical for at least two reasons: (1) they are essential for setting priorities for determining which worker exposures to monitor; and (2) if a group is monitored, the information on rate dependence may help guide the choice of sample averaging time for the OEL and the subsequent interpretation of observed effective exposure data.

Table 2.2 portrays a continuum of exposure rate dependence and appropriate measurement averaging times. If the agent has the potential for high exposures, is fast acting, has poor warning properties, and produces irreversible adverse health effects at the highest potential exposure (on the order of seconds, e.g., hydrogen sulfide at 100 ppm), continuous monitoring with very short averaging times and local alarms is advised. If the effect is reversible and there are low concentration warning properties (e.g., sulfur dioxide), continuous monitoring may not be needed. For agents whose exposure rate dependence is measured in minutes or hours, the short-term exposure limit, excursion limits, and time-weighted average (TWA) concepts are more applicable. Chronic exposure to agents that accumulate over months and years are best characterized by long-term average (LTA) exposures (e.g., lead and silica). However, information about process peaks and resulting short duration exposures may be needed to devise effective control strategies.

TABLE 2.2

The Relationship of Health Effect, Averaging Time, and Exposure Limit

Exposure Duration: Adverse Effect	Appropriate Measurement Averaging Time	Appropriate Occupational Exposure Limit	Example
Seconds: Acute effects	Short-term monitoring and warning	Ceiling	H_2S (Central nervous system effects) N_2 (Asphyxiant)
Hours: Subacute/ chronic effects	Short-term exposure limits and 8-hr TWAs	STEL/TWA[A]	Solvents (Narcosis)
Weeks: Chronic effects	Weekly/ monthly average exposures	TWA[B]	Lead (Long biological half life)
Years: Long-term effects	Annual average exposure	LTA[C]	Silica (Bioaccumulater)

[A]STEL/TWA = Short-term exposure limit/time-weighted average.
[B]TWA = time-weighted average.
[C]LTA = long-term average.

Determination of Homogeneous Exposure Groups

Homogeneous exposure groups are groupings of workers who are expected to have the same or similar exposure profiles or distributions. The HEGs are established by several techniques, but the key element is the assignment of individual workers into groups with similar exposures. Determining homogeneity of exposures to specific agents among worker groups is the goal. However, much like defining fundamental building blocks of matter (molecules, atoms), there are differing levels of complexity/simplicity and also differing levels of practical significance. For purposes of this document, an HEG

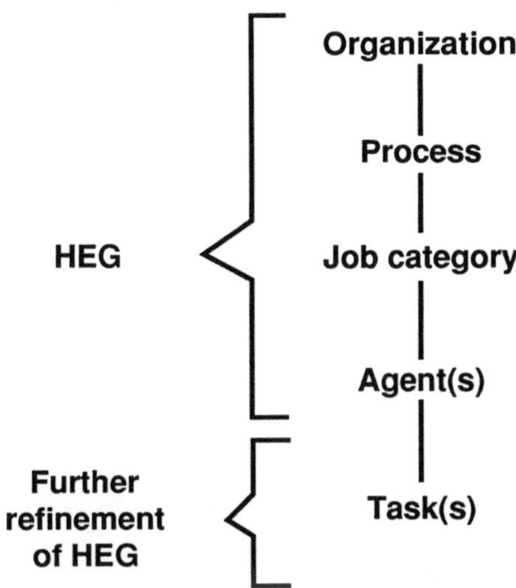

Figure 2.2. Elements of a unique homogeneous exposure group description. HEG = Homogeneous exposure group.

is at least broken down by process, job category, and list of agents for that process/job description. Further refinement (and more homogeneity) may be achieved by breaking HEGs down to process/job/agent/specific task (as illustrated in Figure 2.2).

In general, the HEG approach can be summarized succinctly as a cataloging of job and task descriptions—matching the job task descriptions with actual worker experience and assigning workers to specific HEGs. The job and task catalog must contain specific attributes to be useful. These attributes include associated environmental agents, worker tasks, locations, frequency, and intensity of agent/task interfaces, exposure control measures, and temporal considerations (e.g., summer/winter only, first shift only).

The generation of the job and task catalog starts with the data collected during characterization of the workplace, especially the steps that pertain to the existing job classifications and that pertain to existing task descriptions. Data collected in the preparation steps will vary by workplace. Within a single firm, the available information may vary by location. There are cases where job descriptions and task descriptions compiled for purposes other than industrial hygiene are relatively complete and easily used; conversely, there are cases where the preliminary data from these sources may not be meaningful. In most cases, the existing personnel systems are somewhere between these two extremes. If industrial hygienists use systems other than personnel, documentation linking the exposure group classification with the personnel system is often critical for successful epidemiologic and health surveillance programs.

Starting with a job and task catalog that may contain a number of redundancies and culling out the redundant entries through combination and condensation seems reasonable. This can be achieved by several techniques.

Task-Based Approach Each task associated with a process or subprocess is treated as an entry. Advantages of this type of catalog are the classification of each process/task as a separate entity to be considered with separate subgroupings based on the presence and absence of controls, its associated environmental agents, work-identifying descriptors, frequency of occurrence, and temporal distribution. In essence, the HEGs are created on the basis of workers' task equivalence.

A strictly task-based approach includes only process and task categories with associated agents, and it omits the job description categorization. The disadvantages of this classification are that the number of distinct groupings may become unwieldy, each entry may represent only a fraction of a workday, and it may be difficult to relate results to actual worker exposures. Note that an engineer taking a sample is no different from an operator taking a sample; the act of taking a sample makes them equivalent for that task. The difficulty is proper documentation of the identity of the individual who performed the task.

Job Description–Based Approach The assumption will be made that either the existing personnel job descriptors are sufficiently distinct or a new set of job descriptors based on process, task, and agent similarity is created. Each job description is an entity represented by at least one worker on at least one day. Thus, the job description may involve a number of tasks and agents. A specific worker may undertake different job descriptions from day to day, but the assumption may be reasonably made that the duration is neither excessively short nor infrequent. Advantages of this type of catalog are that each entry is an expression of a distinct workday and expresses the experience of a typical worker on that descriptor. Thus, the physical act of collecting exposure data is simplified to finding the set of workers that fit the descriptor. This classification, especially if it is backed up by employment records, will facilitate epidemiologic studies since work history is automatically compiled. The disadvantage of this classification is that it is highly susceptible to minor procedural and administrative changes (task heterogeneity) in work performed within each job description—changes that are usually not documented. Thus, a periodic review of this HEG may be complicated. If the industrial hygiene job descriptions are not the same as the job description shown on employment records, the two sets of descriptions necessitate specific linking of industrial hygiene and personnel management systems; this may require computer database and communication.

Chemical-Based Approach This method begins with a complete inventory of agents used in an organization or in a workplace. The agents of interest are retrieved from this list, and workers are assigned to uniform exposure categories within the agent category. For example, a firm selects benzene as a priority agent. Each facility in the firm that is working with benzene qualitatively evaluates its workers' exposures to benzene, and homogeneous benzene exposure groups are constructed within facilities and throughout the firm. This fundamentally differs in approach because it starts with the chemical and works upward. The other approaches begin from worker or facility attributes and build down to homogeneous exposures to chemicals. All presented approaches would probably lead to similar groupings for specific agents, but the agent-based approach is weakest in terms of broad classification of mixed exposures. It is most useful for specific agent regulatory compliance or specific agent epidemiologic studies. Agent-based approaches may introduce possible confounding problems in an epidemiologic study if other agent exposures are not identified.

Process/Job/Agent/Task–Based Approach Process/job/agent/task approaches will most likely lead to groupings homogeneous in exposure; however, there may be hundreds or thousands of such HEGs within a facility. This is unwieldy for most practical applications. Industrial hygienists should combine these minuscule HEGs into larger groupings wherever observation and measured exposures justify it.

Data-Analysis Approach In the rare case where large numbers of personal exposure measurements are available, HEGs can be directly determined by a statistical procedure known as analysis of variance (ANOVA). This approach will examine within-worker and between-worker exposure variability and will serve as a basis for separating workers or grouping workers for HEGs. Rarely are sufficient data available for this direct approach. However, ANOVA is a tool that can be used to confirm that an HEG established by professional judgment is indeed composed of homogeneous exposures.

Summary of HEG Approaches Whether one uses a task, job description, process, or chemical approach, the ultimate outcome should be groupings of workers whom the industrial hygienist believes have similar exposure profiles. The HEGs evolve over time—chemicals change, degrees of exposure change, personnel change, etc.—and they must be reviewed and updated regularly. The number of workers involved may require the use of a computer database to track HEGs efficiently.

The sodium chloride production plant example, featured in Appendix I, demonstrates the use of the process job description–based approach to determine HEGs.

Outcomes of the Basic Characterization

The basic characterization produces a set of inventories cataloging tasks, processes, environmental agents, potential for exposure, occupational exposure limits, and workers. Homogeneous exposure groups are determined from this data and each worker is assigned to at least one HEG. These HEGs form the basis for exposure assessment priorities and for exposure monitoring campaigns.

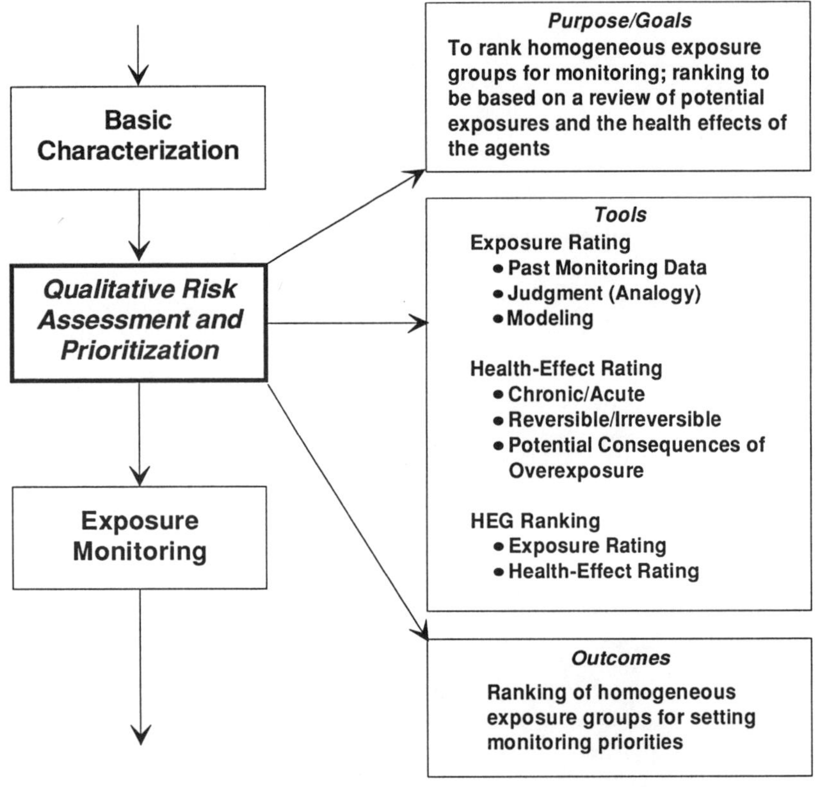

Figure 3.1. Qualitative risk assessment and prioritization component of the exposure assessment strategy

3
Qualitative Risk Assessment and Prioritization

PURPOSE

The outcome of basic characterization is a set of homogeneous exposure groups (HEGs). The task of this chapter is to describe methods for estimating an exposure rating and a health-effect rating for each agent and to establish an HEG priority ranking based on the qualitative risks faced by the workers in each HEG. Design of an efficient and effective monitoring campaign is facilitated if the various HEGs have been assigned priorities for sampling on the basis of the potential risks for each HEG. The qualitative risk assessment process is shown in Figure 3.1

PROFESSIONAL JUDGMENT AND AN OVERVIEW OF TOOLS

The tools of exposure rating, health-effect rating, and HEG ranking are commonly used by industrial hygienists to organize their priorities.

Many qualitative risk ranking schemes can be devised that perform the same function as the generic scheme described below. In fact, whether formalized or not, industrial hygienists often devise schemes that are useful and specific to their workplaces. Every such ranking scheme has professional judgment inherent in it. The industrial hygienist should use or develop a risk ranking scheme that incorporates both components of risk: effective exposure and agent toxicity.

Exposure Rating

Workers within a process are often assigned to an HEG that has homogeneous exposures to a number of agents. Within an HEG and for a specific agent, the workers can be assigned a single

TABLE 3.1
Qualitative Exposure Rating[A]

Category	Description
0 No exposure	No contact with agent
1 Low exposure	Infrequent contact with agent at low concentrations
2 Moderate exposure	Frequent contact with agent at low concentrations or infrequent contact with agent at high concentrations
3 High exposure	Frequent contact with agent at high concentrations
4 Very high exposure	Frequent contact with agent at very high concentrations

[A]Exposure can be by air, dermal, or ingestion routes. Rating should not take into consideration the use of personal protective equipment. Infrequent/frequent definitions depend on the hygienist, the workplace, and the agents; consistency in application is the key, not the definitions themselves (e.g., magnitude of exposure potential may be thought of as multiples or fractions of the applicable occupational exposure limit).

exposure rating. If multiple exposure ratings are needed for a specific agent within an HEG, the HEG must be split into smaller HEGs for which exposures are homogeneous. Table 3.1 shows a generic qualitative exposure rating scheme with five categories of exposure. The industrial hygienist should use past exposure data, exposure modeling, and/or professional judgment to assign exposure ratings.

As an example, assume a chemical processing plant has a single agent. Many plant clerks and secretaries would be Category 0 and some may be Category 1 (if they go into plant areas). Maintenance workers may have a Category 2 exposure rating if they sometimes open pipes carrying the agent and are often in the process area exposed to background levels. Process workers may interact with the agent daily and at higher concentrations. Although the operators wear respirators during peak exposures, their potential exposure ratings should not be weighted or lessened by the use of personal protective equipment

(PPE). Therefore, based on peak exposures, the operators may warrant a Category 3 rating. In this example with one agent, the majority of monitoring would be done on the process workers with some evaluation of the exposure of maintenance workers. The clerks and secretaries would probably not be monitored since their potential exposure is low.

There are several tools that can be used to refine the exposure rating. Three tools are used most often in preliminary exposure assessment:

1. Past exposure data
2. Exposure models
3. Professional judgment (analogy to other exposure situations)

Use of Past Exposure Data If samples have been taken in previous years, the results may provide valuable information, especially if the process has not changed significantly. Even if change has occurred, the monitoring data do provide some basis for estimating potential exposures of workers in HEGs. Adjustments based on professional judgments and models may be made. Plotting past data over time to determine whether the exposure trends are higher or lower may be helpful. If the exposure trends exist, the industrial hygienist may use only the most recent exposure data in the preliminary assessment. If many data are available, some statistical analysis may be appropriate.

Simple Modeling of Exposures Models are used to estimate exposures of populations that for any reason cannot be monitored directly and also as a screening tool to set priorities for more in-depth monitoring studies. Models can be used to estimate both historical exposures that cannot be recreated and possible future exposures in hypothetical situations. Models can be used in cost trade-off calculations when evaluating alternative control technologies. A good model predicts exposure with reasonable accuracy and is a tool that can save time and money in establishing exposure evaluation criteria, specifying control technology, and selecting personal protective equipment. Any occupational exposure model should be used with the ultimate goal of providing the industrial hygienist with additional insight into an exposure problem. Simply by understanding the components of a model, an industrial hygienist's insight about possible exposures is enhanced, even if the model is not perfectly accurate. In the context of this document, simple modeling can be used as an input to an industrial hygienist who must subjectively estimate potential degrees of exposure. Industrial hygienists must keep in mind that most occupational exposure models are crude representations of actual exposure processes and that the model predictions must be interpreted accordingly.

Many of the models commonly used in industrial hygiene provide worst-case estimates of concentration. For example, a model that predicts employee exposures to airborne concentrations of a chemical contaminant is a mathematical relationship linking the source through the airborne pathway to the breathing zone of workers. Sophisticated models recognize that the sources, removal mechanisms, transfer mechanisms, and employee interaction are distributed through time and space.

Mathematical models of simple exposure situations become complicated very quickly. Therefore, the models used to predict occupational exposures often are based on many simplifying assumptions. For example, simple models assume steady state concentrations (e.g., dilution ventilation models) rather than trying to describe time dependence of concentration. Deviations from uniform mixing are accommodated with a single mixing factor. The bottom line is that every simplifying assumption reduces the ability of a model to predict fine details of workplace concentration gradients. As long as the industrial hygienist confirms that the simplifying assumptions tend toward overestimating exposures, then models can be effectively used as one tool for preliminary exposure evaluation.

Another example is the dilution ventilation model, which has been used extensively (readers may wish to refer to the American Conference of Governmental Industrial Hygienists industrial ventilation manual.[*]) The model predicts the steady-state airborne concentration of a solvent when the source rate and dilution ventilation rate are known and assumed constant.

The dilution ventilation model is based on a simple mass balance and in steady state has a form such as the following:

$$C \quad \left(\frac{403 \times SG \times ER \times K \times 10^{6}}{MW \times Q} \right) ppm_{V} \qquad \textbf{(1)}$$

where C = Concentration of solvent in ppm_V
SG = Specific gravity of solvent in g/mL
ER = Evaporation rate of solvent in pints/hr of liquid
K = Safety factor, usually between 3 and 10. The high end of the range should be used if incomplete mixing is present.
MW = Gram molecular weight of solvent
Q = Ventilation rate of workplace in cubic ft per min

Another commonly applied worst-case model estimates saturation concentration. The vapor pressure model is as follows:

$$C \quad \left(\frac{P_{c} \times 10^{6}}{P_{a}} \right) ppm_{V} \qquad \textbf{(2)}$$

where C = Concentration of chemical in ppm_v
P_c = Vapor pressure of chemical
P_a = Ambient barometric pressure

The vapor pressure model establishes an upper bound on the concentration. It does not account for generation rates, ventilation rates, or worker interaction. Worker time/task behavior is often difficult to model, but it is also the key to estimating worker exposure accurately. Therefore, exposure monitoring is the ultimate validation for exposure model predictions.

---

[*] American Conference of Governmental Industrial Hygienists: *Industrial Ventilation—A Manual of Recommended Practice, 20th Edition.* Cincinnati, Ohio: American Conference of Governmental Industrial Hygienists, 1989.

These two simple models—steady-state and saturation concentrations—are useful tools to an experienced industrial hygienist. The model predictions can help guide the industrial hygienist's subjective ratings of degree of exposure for the various HEGs, but they must not be used in a vacuum. Model predictions must be interpreted in light of any available exposure data and the professional judgment of the industrial hygienist.

Health-Effect Rating

Each agent needs to be evaluated subjectively for the severity of the potential health effects and the uncertainty of the data (e.g., animal data versus human epidemiologic evidence). Table 3.2 shows a generic qualitative health-effect rating scheme. The time interval for monitoring is based on the biological nature of the agent being evaluated. An agent with little or no rate-dependence, such as lead, should be distinguished from an acute irritant that causes adverse health effects within minutes of exposure.

For example, if a chemical processing plant has multiple agents and if an agent has no suspected adverse health effects or reversible ones of little concern, that agent may be assigned to Category 0. If reversible effects of concern, such as a primary irritation, are identified, Category 1 classification may be appropriate. Category 2 may include severe, fast-acting irritants; Category 3, corrosive agents or sensitizers that can permanently affect a worker's life, although he may still be able to work. Obviously, if a situation is found that is immediately dangerous to life or health, the situation must be addressed directly and immediately by remediating the problem and providing continuous monitoring with a local alarm.

Chronic agents (including those with toxic and extremely toxic health effects), carcinogens, and teratogens are not as easily rated as the acute health effects given in the example above. Consequently, judgment may play a larger role than with the acutely toxic materials. For example, one might decide to take a conservative approach and rate all suspected or known carcinogens as Category 4 agents. Perhaps another industrial hygienist might decide to rate animal carcinogens

TABLE 3.2
Qualitative Health-Effect Rating

Category	Health Effect
0	Reversible effects of little concern or no known or suspected adverse health effects
1	Reversible health effects of concern
2	Severe, reversible health effects of concern
3	Irreversible health effects of concern
4	Life threatening or disabling injury or illness

with little relevance to humans as Category 2 agents and reserve Category 3 for animal carcinogens and Category 4 for agents with epidemiologic evidence of human cancer.

The industrial hygienist must use professional judgment in applying rating schemes such as these. There is a great deal of uncertainty in the available health-effect data. Conservative decisions should be made where uncertainty exists, erring on the side of health and safety. There are exceptions to any rating scheme; the key is to realize this scheme is for relative ranking/rating of potential health effects of agents. The industrial hygienist should apply professional judgment in order to create a relative rating of agents that seems reasonable and appropriate and is internally consistent.

Qualitative Risk Ranking

Setting HEG priority for evaluation and monitoring is particularly difficult when there are hundreds of agents and dozens of HEGs in a plant. In this setting a qualitative ranking scheme can be used to establish priorities between competing degrees of exposure and potential health effects. There are a number of ways to relate the two rating schemes for a given agent, including descriptive categories and mathematical equations. Whatever type of ranking scheme is used, the purpose is to scale the qualitative relative risks of HEGs exposed to agents between a point of little or no risk to the highest risk of health effects.

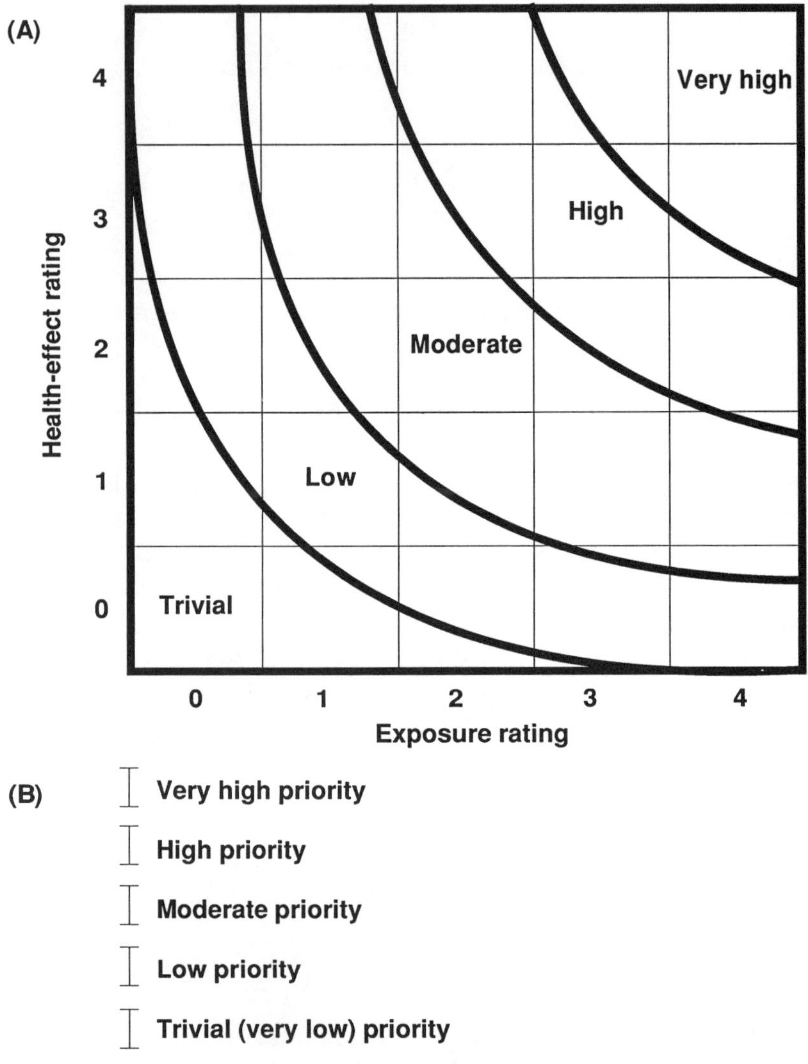

Figure 3.2. A qualitative risk ranking scheme. A = Priority rankings based on qualitative risk assessment; B = Ranking scale for homogeneous exposure groups.

One form of a qualitative risk ranking scheme is shown in Figure 3.2A. It is obvious that an HEG that has a health-effect rating of 4 and a degree of exposure of 4 would get top priority for a monitoring campaign, and an HEG that has both a health-effect and an exposure category of 0 would require no monitoring. The value of a priority risk ranking scheme and the inherent need for professional judgment come into play in ordering the risks in the middle ranges.

In Figure 3.2A, elliptical contours were chosen as a first cut in establishing a prioritization of the HEGs' relative risk. Once the qualitative risks for the HEGs have been determined and scaled on Figure 3.2B, they should be reviewed to see if they appear reasonable. Are there HEGs ranked too low or too high in terms of monitoring priority? For example, a carcinogen, though rated with a low exposure, might be moved from a lower ranking to a higher ranking.

The industrial hygienist should use professional judgment to adjust the ranking of risks and should be alert as to which risks can change an HEG's priority rank (e.g., health effects, regulation, risk communication, etc.). Some additional factors that should be considered in ranking HEGs include (1) frequency of exposures, (2) number of workers in the various HEGs, and (3) emphasis on monitoring groups close to an occupational exposure limit (those groups clearly overexposed should be dealt with immediately using control measures). A qualitative risk ranking scheme such as this is only a tool to assist an industrial hygienist in establishing priorities for monitoring. In spite of the emphasis on finding the highest risk agents, some resources should be allocated to lower risk categories in order to gather baseline data and to confirm subjective ratings of lower exposures.

Additional refinement in ranking results may be achieved by considering the concentration specified in an exposure guideline. For example, when several substances have the same general health effect and qualitative risk ratings, the highest priority may be given to the agent that will be the most difficult to control with available technology. One first-cut analysis for making such determinations is to calculate the ratio of the occupational exposure limits (OEL) to the saturation concentration (reciprocal is known as the vapor hazard). The lowest ratio (OEL/worst-case level) may provide guidance on which agent will be the most difficult to control. High rankings should always be given to the substances most likely to be at or above exposure guideline levels. Higher rankings should be given to exposures controlled by personal protective equipment than to exposures controlled by process and plant design.

Outcomes of the Exposure Ranking

The list of HEGs ranked by relative risk is the primary outcome of this step and is the primary input into the design of a monitoring campaign.

Those groups with risks rated negligible or low will probably not be monitored. Groups with moderate risk will be monitored to some degree, and high and very high risk ratings will receive the most monitoring. HEGs facing exposures to multiple agents may require more consideration by the industrial hygienist. If the agents potentially act as additive or synergistic toxic agents, each agent assigned to an HEG may need to be monitored. However, if their toxic effects are independent, individual rankings for the agents may guide monitoring priorities. When actual exposure data are acquired, the exposure ratings for each HEG should be reviewed and updated as appropriate. This is insured (as shown in Figure 1.2) by the periodic reevaluation.

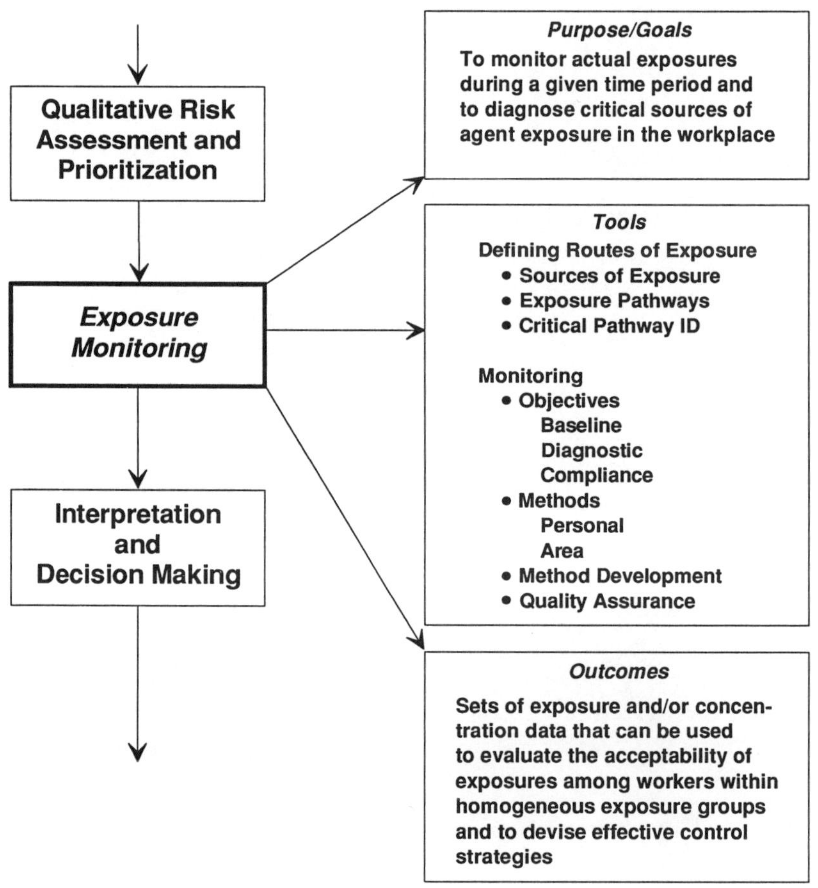

Figure 4.1. Monitoring component of the exposure assessment strategy

4
Monitoring

PURPOSE

Once a decision to monitor exposures among workers in homogeneous exposure groups (HEGs) is made, the focus must shift toward devising a monitoring campaign that provides the most information possible per sample. This chapter covers the elements of a sound monitoring strategy and campaign and addresses, in the context of the subsequent interpretation and decision making, the vexing issue of the number of samples required.

Monitoring is the measurement of exposures or contaminant concentrations during a given time period. Monitoring results can be used for quantitative estimation of exposures for individuals and groups of workers and for identification of sources of potential exposure problems. The type of monitoring selected for a campaign is critically linked to the question an industrial hygienist needs to answer. To assist industrial hygienists in selecting appropriate monitoring techniques, an exposure pathway model is presented. The monitoring step is illustrated in Figure 4.1

TOOLS

The tools for a monitoring strategy and campaign include (1) a model for exposure pathways; (2) a clear statement of monitoring objectives (including baseline, diagnostic, and compliance monitoring); and (3) an array of monitoring methods (including area and personal monitoring).

The industrial hygienist should start with an exposure pathway model and subjectively determine the potential critical exposure pathways. Next, the industrial hygienist should use a mix of the proper monitoring objective (baseline, diagnostic, or compliance) with the proper monitoring methods (personal or area) to evaluate the critical pathways. The selection of monitoring techniques requires judgment and insight into the exposure pathways. Too often industrial hygienists "hang a pump" on a worker for a full shift and later return to collect the charcoal tube. Careful thought prior to monitoring pays large dividends in cost effectiveness. Monitoring should be directed to assessment of the critical exposure pathway. Blindly hanging pumps is both inefficient and ineffective for evaluating and controlling critical exposures.

Exposure Pathway Modeling

Often an industrial hygienist's first instinct is to monitor an air concentration or exposure, but the complexity of exposure scenarios requires consideration of all possible exposure pathways before a monitoring technique is selected. The exposure pathway model in Figure 4.2 highlights the potential pathways leading from process to worker and provides a framework for evaluating pathways of each potential environmental stressor. Once an agent has become airborne, comes into contact with the skin, or settles onto work surfaces, exposure

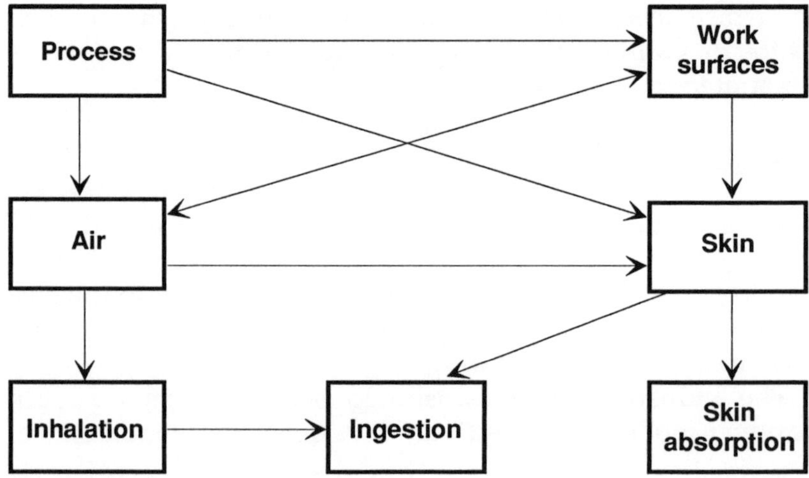

Figure 4.2. Exposure pathway model

can occur via inhalation, skin absorption, or ingestion. Such a model should be considered for each workplace exposure situation since the critical routes of exposure are the fundamental bases for selecting appropriate monitoring methods. Without considering all pathways, an industrial hygienist may sometimes overlook the critical exposure pathway.

Using the exposure pathway model to evaluate each potential route of exposure, an industrial hygienist subjectively selects the critical exposure pathways and the critical points in those pathways. Once the critical pathways have been determined, tailored monitoring strategy can be devised to evaluate exposures on these critical pathways. A high vapor pressure solvent may be primarily evaluated for inhalation exposure unless there is also significant skin contact (either in bulk or vapor form); then, biological monitoring may be needed. On the other hand, if an agent has very low volatility, the critical pathway may be skin absorption and/or ingestion. A wipe testing program may be used to evaluate contamination, and a biological monitoring program, if available, may be used to monitor dose.

Inhalation Exposure Inhalation exposure is often the most significant route of workplace exposure. Agents can directly enter the air from the process, or the pathway can lead from the process to work surfaces and then to the air. With lower vapor pressure materials, there may be a dynamic equilibrium between the plant air and the work surfaces.

Skin Absorption Skin absorption (dermal) exposures can also represent an important route of exposure for some agents, particularly those that pass through the skin and bioaccumulate and those that readily absorb through the skin in immediately toxic amounts. This exposure pathway generally involves direct skin contact either with process agents or with work surfaces that have been contaminated with process agents.

Oral Ingestion Oral ingestion of agents is usually a relatively minor route of exposure in the workplace, particularly when employees follow a reasonable level of personal hygiene and eating areas do not have significant levels of agent contamination. However, this may be an important exposure route for some chemicals (e.g., lead) that bioaccumulate. Also, mucociliary clearance may cause airborne agent exposures to result in ingestion as cleared mucus is swallowed and digested.

Monitoring Objectives

Monitoring objectives can be broadly grouped into three categories: baseline, diagnostic, and compliance.

- Baseline: What is the range and distribution of exposures for defined HEGs?
- Diagnostic: What are the sources and tasks that pose the greatest potential exposures in a workplace?
- Compliance: Is this workplace, in regard to its exposures, in compliance with governmental standards?

Baseline Monitoring Baseline monitoring is performed to evaluate the range or distribution of exposures among specified HEGs. The baseline is the primary database used for determining acceptability of exposures and the need for diagnostic sampling and additional controls.

Baseline monitoring, if properly documented, can also be useful for later epidemiologic studies.

Diagnostic Monitoring Diagnostic monitoring is performed for specific, special case exposure evaluations. Its main goal is to identify the predominant sources and tasks causing exposures. The results serve as a basis for an industrial hygienist to devise the most appropriate and efficient control strategies for the identified high exposure sources. If the source is fixed in installed equipment, area monitoring techniques are appropriate. If the source moves with a worker or is dependent on individual work practices, personal monitoring techniques are appropriate. In most cases, sample averaging times should be adjusted to reflect process cycle times rather than health-effect averaging times.

Compliance Monitoring Compliance monitoring is performed to make formal comparisons with organizational guidelines and governmental standards for acceptable exposures. The sampling strategy for evaluating compliance is usually a worst-case or "most exposed worker" approach, and its execution has been described in many government standards and guidelines.

Monitoring Methods

There are two broad groupings of monitoring methods. They differ in their objectives and in the techniques used for monitoring.

- Personal: To what and at what concentration levels is a worker exposed during a full shift or during a task? This method focuses on worker exposure and must use sampling equipment that is unobtrusive so that normal work procedures can be continued during sampling.
- Area: What is the general background concentration level experienced by most workers in the area? This method focuses on plant conditions and can use permanently installed monitoring equipment as well as portable methods.

Personal Monitoring The goal of personal monitoring is to characterize the dose of an environmental stressor that a worker receives during the averaging time of interest. The dose is seldom measurable, so surrogates for the dose must be measured (e.g., effective exposure). The most common surrogate is worker breathing zone concentration (i.e., the average concentration of an agent near the worker's nose during the time interval of interest). Industrial hygienists should never forget that the measured concentration represents potential exposure; that the effective exposure is lower when personal protective equipment is worn; and that effective exposure is a surrogate for the true quantity of interest, dose.

Personal monitoring can be performed to evaluate three critical exposure pathways: inhalation, ingestion, and skin absorption.

Inhalation exposures (concentrations) are often evaluated by the common personal air sample collected in a worker's breathing zone. These time-weighted average (TWA) exposures represent the average concentration of a contaminant potentially inhaled by a worker. Two common sample averaging times are full shift (8 hr) and 15 min, both of which correspond to common TWA exposure guidelines (e.g., threshold limit value–time-weighted average [TLV®–TWA] and threshold limit value–short-term exposure limit [TLV–STEL]).

Other sample averaging times may be suitable in light of health-effect data. When brief peak exposures cause health effects, shorter sample times are meaningful. When a substance accumulates in the body with little or no removal or transformation, a long-term TWA exposure over months or years may be appropriate. If the agent is a physical agent that has a path from process to worker via the air, other appropriate techniques may be used; for example, noise exposures can be evaluated using sound pressure level instruments and thermal stress can be evaluated by wet-bulb globe temperature instruments.

The industrial hygienist can monitor air concentrations over full shifts to estimate TWA exposures, can monitor shorter peaks when health-effect properties indicate their importance, or can measure exposures during high exposure tasks to estimate the potential contribution of these tasks to a worker's total exposure profile.

Task sampling is often appropriate when evaluating whether short-term exposures are within short-term exposure guidelines or ceiling limits. Task sampling, in some ways a diagnostic tool, is also a very efficient technique for understanding an HEG exposure profile and devising control techniques to reduce overall exposure. Sometimes task samples are used in conjunction with area background samples to calculate an estimate of a full-shift average exposure.

Full-shift sampling, while relatively simple, has the disadvantage that peak exposure information is usually lost. There is no way to identify which tasks led to the peaks. A proper mix of full-shift and short-term task sampling is usually the best technique for evaluating personnel exposures.

Skin absorption is a difficult pathway to evaluate. Available tools include patch testing and skin washes to estimate amounts of a substance on a worker's skin.

Biological monitoring should be considered to evaluate exposures occurring through all three pathways—inhalation, ingestion, and skin absorption. A biological monitoring result will reflect integrated exposure from all three pathways and in some cases is a closer indication of dose than measures of breathing zone concentration such as an air sample. Some limitations in using biological monitoring include individual physiological variability, lack of background data, and unavailability of validated guidelines and standards. Without this information, there is little utility in evaluating exhaled breath, blood, or urine samples from employees. Biological monitoring should be performed only with validated methods so that obtained results can be linked in a meaningful way to existing occupational exposure limits (OELs) and potential health effects.

When available, toxicokinetic models and validated biological monitoring techniques permit the collection of biological specimens and the use of concentrations of the agents (or their metabolites) to estimate doses to critical organs (e.g., the liver). Even when toxicokinetic information and validated methods are available, the evaluation of biological monitoring results may be very difficult, particularly if an adequate control group is not maintained and simultaneously evaluated. In practice, biological monitoring is usually performed in conjunction with a medical department or a consulting occupational health physician and is rarely the sole responsibility of the industrial hygienist.

Area Monitoring Area air monitoring is usually performed to evaluate background concentration levels being experienced by most workers in an area or to provide real-time evaluation of acutely hazardous substances (e.g., continuous air monitors with alarm-warning systems). Area monitoring results can be used in time trend analysis to detect seasonal variation, process cycling, or changes in efficiency of ventilation controls. For groups of workers with no high exposure tasks, the area background concentration may serve as a reasonable estimate of full-shift exposures. However, area monitoring should not be viewed as a substitute for personal monitoring since worker tasks often influence breathing zone concentrations.

Area monitoring also includes investigation of surface contamination by wipe testing methods. Although not a direct measure of exposure, wipe tests are useful for tracking levels of contamination, particularly for substances readily absorbed via the dermal route. Wipe tests are a quantitative means for determining adverse trends in work practices and housekeeping procedures (e.g., cleanliness of lunchroom areas). They also can identify deficiencies in maintenance or operation of exposure control systems such as safety hoods.

The Big Two: Duration and Number of Samples

The final step prior to any monitoring campaign should be to determine how long to collect each sample and how many of those samples are needed. There are no perfect rules. Sometimes a review of data indicates that the wrong choices were made and resampling is needed.

Averaging Time of Sampling Devices For validated biological monitoring methods, the appropriate sampling times are often well-established since the toxicokinetics of these substances are well understood before a validated method is developed. Personal air monitoring, however, has often been performed without considering the toxicokinetics of the monitored agents. Samples are often taken for 8 hr or for 15 min simply because the standards are written with those averaging times. But the 8-hr TWA and 15-min STEL were developed for convenience and have little or no relationship to the toxicokinetics of given stressors.

Industrial hygienists should review all available health-effect data for an agent and try to match the relevant toxicokinetic rates to sample averaging times. For example, a substance that accumulates in the body with little removal (e.g., silica) can best be described by long-term average exposure. On the other hand, a substance that acts as an irritant in a time frame of seconds should be monitored for peaks lasting seconds. If such an irritant were monitored for 15-min periods, the peaks causing irritation could be averaged out and remain undetected and, therefore, not controlled.

Most environmental agents fall between these extremes. Many stressors have both acute effects at high exposures and chronic effects at lower exposures. When this occurs, a two-tiered monitoring strategy should be used to collect samples with both a short and a long averaging time. All OELs should be established with a specific sample averaging time linked to the biological rates of an agent. Industrial hygienists are encouraged to do their health effects "homework" before taking a sample. Direct comparison of monitoring results with standards established with specified averaging times should only be performed where the sampler averaging time is close to the specified averaging period.

Likewise, for diagnostic sampling, the averaging time should be selected to represent process cycle times. Using a longer averaging time will simply average out the very data required for problem identification.

Number of Samples Required Determining the number of samples required has been a challenge for industrial hygienists for decades. Before statistical sampling methods became common, the answer was simple: One should take as many samples as needed to feel confident of one's professional judgment in a given situation (i.e., enough to confirm or contradict the subjective model that the industrial hygienist established based on a qualitative assessment). For some industrial hygienists, that meant 10 exposure samples; for others, it meant 1.

Even today, professional judgment is still the predominant driving force behind the selection of the number of exposures monitored. For example, setting up HEGs and monitoring priorities requires an industrial hygienist to feel comfortable that no samples are required in certain HEGs yet still confident that their exposures in those HEGs are acceptably low or nonexistent.

In the realm of statistical sampling strategies, however, sufficient numbers of random samples must be taken to perform statistical tests. Samples are often taken to estimate population parameters of an exposure distribution. The

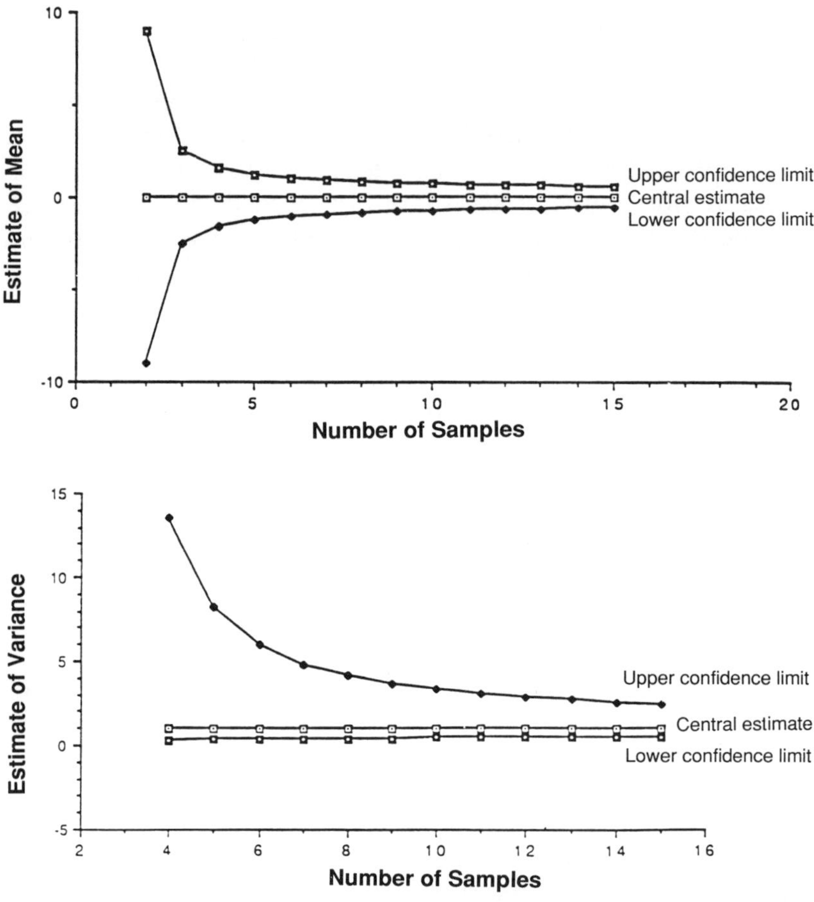

Figure 4.3. Effect of sample size on estimating population mean and standard deviation of unit normal distribution

parameters usually estimated are the arithmetic mean (measure of central tendency) and the standard deviation (measure of dispersion about the mean value). A review of statistical sampling theory reveals that a point of diminishing returns exists. That is, a certain minimum number of samples is needed to estimate the parameters with acceptably small variance (uncertainty). More samples (after a minimum baseline) often do not provide much additional information (as illustrated in Figure 4.3). These figures were developed using a *t*-table and an assumed population distribution that is unit normal. (Discussions of descriptive statistics and probability plotting and of arithmetic mean tests appear in Appendixes II and III.)

As Figure 4.3 also shows, a plateau is reached in estimating these two parameters after about 6 to 10 samples. Fewer than 6 samples leave a great deal of uncertainty about each parameter (and hence in the exposure distribution). More than 10 samples provide additional refinement in estimates, but the marginal improvement is small considering that cost per sample is essentially constant. The plateaus indicate that at least 6 random samples should be taken for each HEG monitored, unless measured exposures are much less than standard (e.g., <1/10 standard) or much greater than standard (e.g., several times greater)—situations requiring fewer samples for a decision. A reasonable approximation to an exposure distribution is usually possible with about 10 samples; however, for rigorous goodness-of-fit testing for a distribution, 30 samples or more are often needed.

One point must be clear, however: These 6 (or more) samples should be *randomly* acquired from the defined homogeneous population in order to perform statistical tests later. All commonly used statistical tests assume random sampling. For a given homogeneous exposure group, the 6 or more samples should be randomly assigned to different workers on different days over the time period of interest (e.g., 1 yr). The population from which random samples are taken should be as homogeneous as possible; that is, if a population can be further categorized (stratified) based on tasks or systematic process features (e.g., every Monday new chemicals are unloaded from tank cars), stratifying before random sampling may be appropriate. Data to be analyzed together must be of the same or very similar averaging times (i.e., 15-min TWAs should not be mixed with 8-hr TWAs in performing a statistical analysis).

On days when sampling takes place, a worker (or workers) from the HEG (or stratified HEG) should be randomly selected for monitoring. Other samples can be taken on monitoring days for diagnostic purposes for better understanding of the process and exposures; however, these samples should not be statistically analyzed with random samples that are expressly collected to estimate the exposure distribution. If the samples have not been randomly collected, statistical tests should not be applied. (Professional judgment would be the only tool available for evaluation if samples are nonrandom). If a good faith effort has been made to try to select days/people at random, industrial hygienists may apply statistical tests so long as they feel comfortable that there is no systematic problem with how data were collected.

OUTCOMES

In this chapter, the monitoring tools have been described and the biological and statistical rationale for determining sample averaging times and the number of samples have been discussed briefly. The outcome of this step is that a set of samples (exposure data) that are linked to the appropriate critical exposure pathways and are suited to statistical analysis should emerge. The next chapter describes decision-making tools that can be applied to these exposure data in order to assist the industrial hygienist in evaluating whether exposures (and, therefore, risks) are acceptably low.

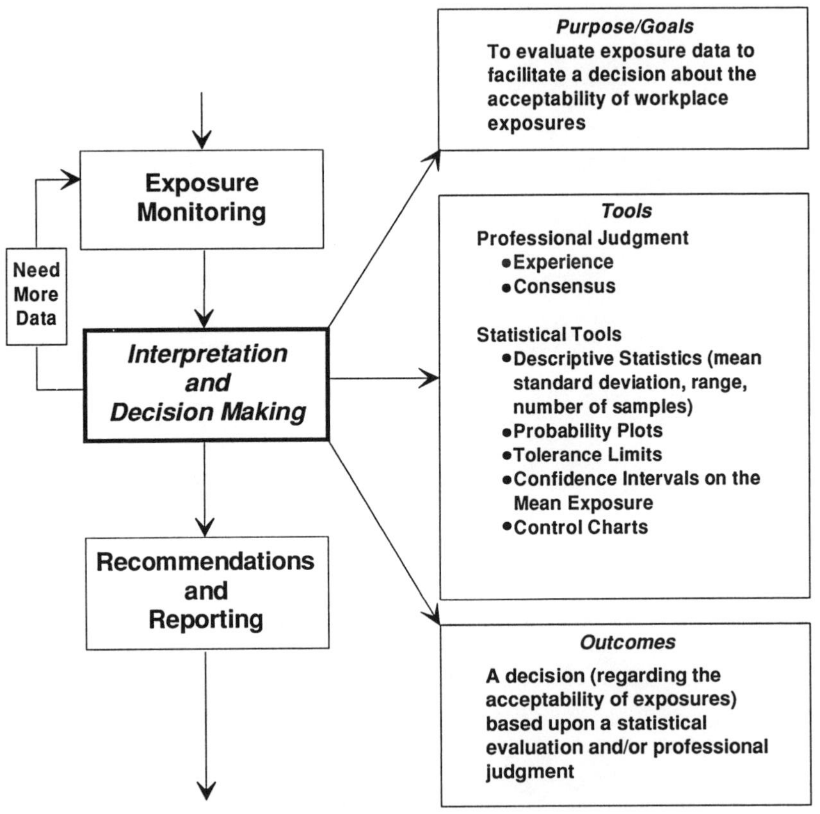

Figure 5.1. Interpretation and decision-making component of the exposure assessment strategy

5
Interpretation and Decision Making

Purpose

The task at hand is to interpret the exposure data representing each homogeneous exposure group (HEG) and to decide whether the exposures demonstrate an acceptable work environment. The two aspects of the decision-making process are quantitative and qualitative: statistical analysis and professional judgment, as shown in Figure 5.1. Even the quantitative aspect depends on the qualitative assumptions underlying the statistical analysis—professional judgment is required both to choose which exposures to monitor and later to interpret the meaning.

Decisions, Decisions—An Overview

There are only three possible decisions when analyzing acceptability of occupational exposure data.

1. The exposures are acceptably low.
2. The exposures are too high and corrective action is necessary.
3. There are insufficient data to make a decision. This includes cases where results of quality control techniques (spikes) and/or professional judgment have shown the data to be erroneous.

These three decisions can be based upon statistical tests, professional judgment, or a combination of both; the sampling strategy this document outlines leaves the industrial hygienist monitoring and/or controlling exposures until acceptable exposures have been demonstrated. Remedial actions may be required if the exposures are found unacceptable, and subsequent additional monitoring will be needed to determine whether the exposures are acceptable after such action has been taken. If changes have been made and new data collected, they should be analyzed independently of the first set of data. Data pooled for analysis must represent a single HEG. The statistical tools discussed in this chapter are reviewed in greater detail in Appendixes II through VI.

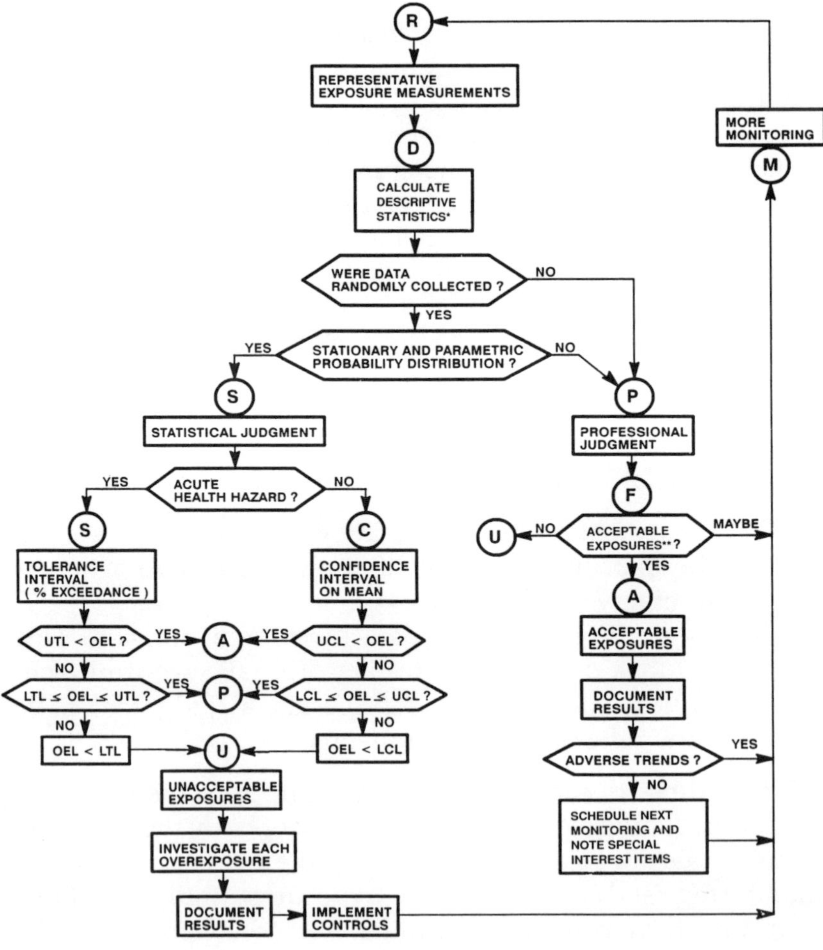

Figure 5.2. One decision-making scheme. *See Figure 5.3. **See Figure 5.4.

Professional Judgment

The description in the diagrammed interpretation scheme (Figure 5.2) includes a professional judgment loop. Professional judgment (in this context) refers to the integration of all that is known about a process in addition to the current monitoring results, culminating in an evaluation of the acceptability of exposures.

No Measured Overexposures If all exposures are less than an exposure guideline and the exposures seem representative of the population of exposures, an industrial hygienist may decide that the workplace exposures are acceptable. If the industrial hygienist is uncertain about the representativeness of the samples, more exposures should be monitored before making a decision. Simple nonparametric calculations can be used to estimate the statistical confidence supporting the decision that the exposures are acceptable. (Appendix VI discusses nonparametric statistical methods in more detail.)

One or More Measured Overexposures If any exposure measurement is over the guideline, an investigation is needed. Further, diagnostic sampling may be needed to identify the source of the overexposure. If the days or tasks with exposures above the exposure guideline can be predicted, then controls should be focused on those specific days or tasks. If the exposures above the guideline cannot be predicted, then consideration should be given to the incidence of the overexposures suggested by the database. More sampling is often needed before an industrial hygienist can make a professional judgment. If the investigation indicates that the overexposures seem representative of the population of exposures, the industrial hygienist should institute immediate process changes or control improvements. Once the improvements are made, a new data set should be collected. However, if an overexposure is not representative of the population of exposures (e.g., an unusual spill or accident caused the excursion), an industrial hygienist may subjectively determine that normal exposure levels are acceptable. If overexposures are occurring, a more frequent periodic monitoring schedule may be appropriate.

Use of Parametric Statistics The first step in the selection of any statistical tool is to determine whether the data meet population distribution assumptions. Plotting the data points is a very effective way of examining the data set for outliers and patterns. Plotting the histogram of the exposure data or graphing the probability plot on the appropriate graph paper (e.g., lognormal probability paper) can answer the following questions. Does the assumed distribution correctly fit the data set? Is there a straight line that seems to fit the data plotted on the probability graph paper?

If the fit to the assumed distribution seems adequate (a subjective decision), one can proceed with a parametric statistical analysis. If not, then nonparametric analysis may still be helpful. Formal goodness-of-fit tests are usually ineffective until sample sizes approach 30 or more random samples because of low statistical power.

The parameters of the assumed distribution can be estimated directly off a fitted line on a probability plot. For example, assuming lognormality, the median (or geometric mean) is the value of the 50th percentile, and the geometric standard deviation can be estimated by dividing the 84th percentile by the median estimate. Reading from plots is preferable to calculating these parameter estimates when sample size is small.

Defining the Acceptable Workplace To assist in decision making when statistical assumptions have been met, two general categories of tests are recommended.

1. An arithmetic mean exposure test (discussed later in this chapter) for stressors or environmental agents posing a chronic risk
2. A percent exceedance test (discussed later in this chapter) for stressors or environmental agents posing an acute risk

Statistical Analysis

The statistical tools are powerful only if their theoretical bases and limitations are understood by the person using them. The practicing industrial hygienist must know statistical fundamentals well. This means knowing and understanding proper use of a confidence interval, a tolerance interval, and a hypothesis test.

The industrial hygienist should understand some basic probability distributions and what they actually represent. Without knowing these fundamentals, the industrial hygienist cannot evaluate the basic assumptions underlying statistical tools and may end up using them incorrectly.

There are two major types of statistics used in industrial hygiene: parametric and nonparametric. Parametric statistics are the most commonly practiced, but they also require the most assumptions. The parametric methods hinge upon one key assumption—that the true shape of the probability distribution of exposures is known. This assumption is often uncertain, and usually insufficient data are available to verify the assumed distribution shape statistically.

The second major category, nonparametric statistics, is just the opposite. These often assume no distributional shape to the data and tend to focus more on robust measures like the median or other percentiles. Robust measures such as these are less sensitive to outliers and spurious data. Thus, nonparametric statistics that use these robust measures are largely independent of the distribution of individual observations. As attractive as the nonparametric statistics are for minimizing assumptions, these methods tend to have lower statistical power and usually require many more samples for statistically based decision making.

When using statistical estimation tools and holding the confidence level constant, the tools rank as follows. The mean is estimated by a few samples (6 to 11). The variance and the 90th percentile require a few more (20 to 50). Accurate calculation of the upper tolerance limit for a 95th percentile can require hundreds of samples. From another perspective, if a decision must be made with a few samples (e.g., 13), the level of confidence is highest for the estimate of the mean, lower for the estimate of variance, and lowest for estimates of large percentiles.

Critical Assumptions Before proceeding to the performance of any statistical approach, two questions must be asked and answered by the industrial hygienist.

1. Were the data randomly sampled? Within the homogeneous exposure group, was there a systematic selection of workers, dates, or times for monitoring? If so, then statistical analysis is inappropriate. In obviously nonrandom cases, simple descriptive statistics can be calculated, but ultimate decisions should be made on the basis of professional judgment. If a good faith effort has been made to acquire a random sample, the industrial hygienist may choose to use statistical methods. Whatever the type of samples, good industrial hygiene practice ensures that any exposure greater than an exposure guideline should be investigated and evaluated.
2. Is the population distribution of exposures changing? Random samples are collected to estimate the parameters of a stationary population distribution. If the population distribution is changing over the period of random sampling (e.g., underlying distribution is nonstationary), then only calculations of simple descriptive statistics and decision making on the basis of professional judgment are recommended. Changes in process materials, physical process parameters, work practices, and performances of engineering controls are common examples of changes that can affect the population distribution of exposures.

Although not an assumption *per se,* a third criterion must be met: The data to be analyzed must be of the same or very similar averaging time (e.g., one should not statistically analyze 8-hr time-weighted average (TWA) data with 15-min TWA data). Mixing of data of different averaging times makes estimates of variance inaccurate and precludes use of most common statistical tools.

Descriptive Statistics Starting the analysis of exposure data by calculating a sample arithmetic mean and standard deviation, sample median, range, maximum and minimum exposures, actual fraction of exposures over the occupational exposure limit (OEL), and the sample size is useful. These simple descriptive statistics provide the summary information that can guide an experienced industrial hygienist through a subjective evaluation of the exposures in a workplace.

Data representative of an HEG should be plotted on probability paper. The two types of paper most commonly used in industrial hygiene are normal-probability paper and log-probability paper. If the data plot as a straight line on either chart, that is strong evidence that the data are adequately fit by a parametric distribution—normal or lognormal, respectively. If this is the case, then percent exceedance levels can be estimated directly from the plotted data—a real asset when one uses professional judgment about workplace exposures.

Mean Exposure (Confidence Interval) For chronic-acting substances such as lead or silica, long-term average (LTA) exposures are the most relevant indexes of dose. This results because dose is cumulative in these cases. Knowledge of high or low acute exposures is less relevant—the arithmetic average summarizes the total mass absorbed by a person. Statistical tests have been developed that evaluate whether an arithmetic workplace mean exposure is less than an exposure guideline. Usually the exposure guidelines used for comparison will not be a permissible exposure limit (PEL) or a threshold limit value (TLV®) but will be a long-term average limit. If a standard 8-hr PEL is used for comparison, a large fraction of exposures could still be over the PEL while the mean exposure remains below the guideline. This is contrary to the intended interpretation of most 8-hr PELs. A typical LTA limit may be one-third of an 8-hr PEL. If an OEL was set on the basis of a risk assessment using the OEL as an average value over long periods of time, using the OEL without correction may be appropriate. LTAs should be set by exposure guideline authoritative bodies for those chemicals that primarily pose health risks from chronic exposures; to date, there are few specified LTAs.

The mechanics of using a mean exposure test are described in Appendix III.

Percent Exceedance (Tolerance Interval) Occupational exposure guidelines are established so that—with the highest certainty permitted by available data—most workers will not suffer health effects if exposed at the guideline level day after day for a working lifetime. Implicit in that description is the possibility that a small fraction may indeed experience health effects at or below the guideline level. This is one reason why all exposures should be kept as far below guideline levels as reasonably achievable. Because of the inherent variability of workplace concentrations, guaranteeing that all exposures are below a guideline is impossible. Demonstrating statistically that no more than a given percentage are greater than a standard, however, is possible. This notion is the basis for a percent exceedance test. The most common test of this type now in use in industrial hygiene is the tolerance limit approach. This approach will permit a test to find whether one can have, for example, 95% confidence that no more than 5% of the exposures exceed the standard. An industrial hygienist can select whatever percentages are appropriate in light of an agent's toxicologic effects, warning properties, and general toxicologic uncertainty.

There are both parametric and nonparametric versions of the tolerance limit approach (the nonparametric requires many more samples).

The mechanics of using a tolerance limit approach are described in Appendix IV.

Trend Analysis (Control Charts) Once workplace exposures are controlled to an acceptable level, some analysis of trend of exposure level over time may be appropriate. This time trend may be reviewed periodically to assure that control is being maintained.

Data from current and past reports may be summarized as mean exposures and confidence intervals around the mean exposure. These means and confidence intervals can be plotted versus date of sampling, and any qualitative trends in exposure levels should be evaluated. If exposures seem to be trending higher over time, more frequent monitoring campaigns may be needed. If trends appear stable or exposure levels seem to be decreasing, less frequent monitoring may be acceptable, freeing monitoring resources for higher priority HEGs.

Some industrial hygienists apply formal control chart methods to assure that current exposures are in a state of statistical control relative to past observed exposure data. Use of control charts can help evaluate whether the distribution of exposures is stationary or is changing over time. Control charts were primarily developed to evaluate samples of "widgets" in production facilities, all of which needed to meet the same tolerances. Modern factories with properly designed control systems should meet the tolerances specified by the OELs selected by plant management. Control charting is the proven tool for this task.

The basic mechanics of control chart analysis are described in Appendix V.

Putting It Together—One Sample Decision Tree

Figures 5.2–5.4 provide a decision flow diagram of the concepts of this chapter. The process starts or recycles at the point labeled R in Figure 5.2. At this point it is assumed that a valid HEG has been defined, that the agents of interest have been prioritized for monitoring, and that representative random samples have been selected for monitoring. This means that all circumstances in the workplace exposure process are equally likely to be picked (seasons, shifts, days of the week, etc.). Finally, the proper monitoring and analytical methods have been used to ensure data integrity. Generally, a limit of detection at or below OEL/10 is preferable.

At point D of Figure 5.2, the reader is encouraged to review Appendix II. The main point of interest is that descriptive statistics (mean, median, mode, variance, geometric mean, geometric standard deviation, minimum, maximum, and range) are all calculable for any set of data. No assumptions are required at this step. The details of the descriptive statistics process are expanded in Figure 5.3. In any ongoing industrial hygiene surveillance program, maintaining control charts for the purpose of detecting trends over time that might indicate deterioration in installed engineering controls or in worker compliance with safety and health procedures is important.

If these assumptions are met, then the data are plotted on probability paper to see if they lie along a straight line. If linear,

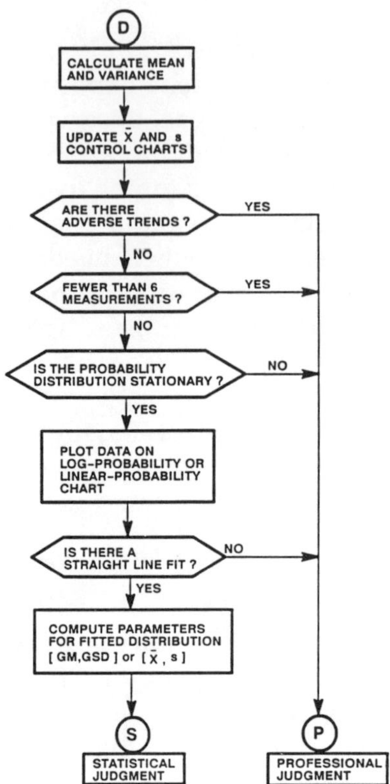

Figure 5.3. Calculating descriptive statistics

then parametric statistics may be used to strengthen the conclusions about the workplace. Most industrial hygiene data are lognormally distributed and would be described by two parameters—geometric mean and geometric standard deviation. From these parameters one can calculate the mean exposure using Equation II.7, Appendix II.

At point S of Figure 5.3, the decision flow returns to Figure 5.2. Procedures from Appendix III are used to calculate the confidence interval on the mean and those from Appendix IV to calculate the tolerance limits on the population of exposures. From either type of analysis, a determination of acceptability of overall exposures (mean or population distribution) is made. The point is made at the bottom of Figure 5.2 that each overexposure is to be investigated by review of field notes and interviews with worker(s) represented by the sample.

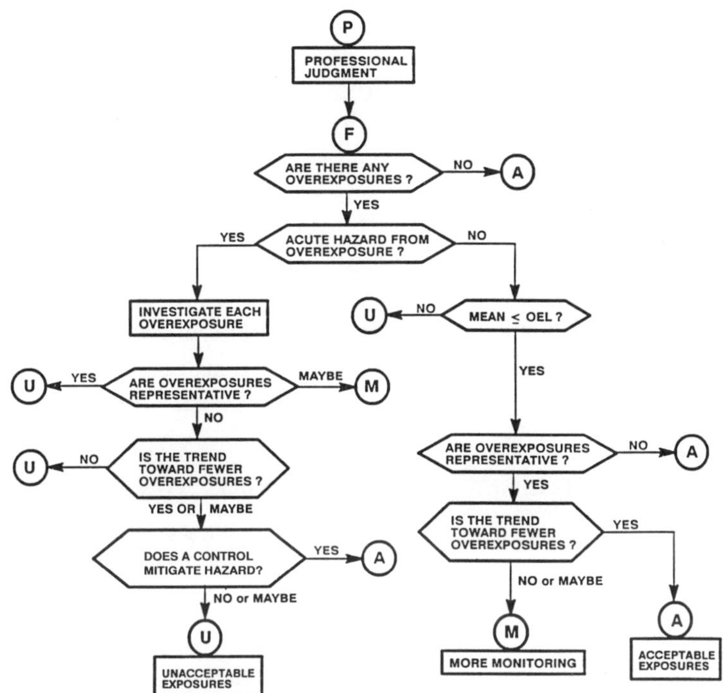

Figure 5.4. Professional judgment

If any of the assumptions justifying the use of parametric statistics are violated or if the parametric tests are inconclusive, the decision flow goes to point P. The professional judgment process is illustrated in further detail in Figure 5.4. Point F needs a little clarification. If nine or more samples are all less than the OEL, the decision to call the workplace acceptable may be fully justified from a nonparametric statistical basis (Appendix VI). However, if the measured exposures are all less than OEL/10, as few as six samples can be used to justify this decision.

Below the fork in Figure 5.4, the kinds of issues to be considered for agents posing ongoing acute risk (left fork) and for those posing a lifetime risk and having negligible dose-rate effect (right fork) are illustrated. These are meant to identify the minimal essential factors a professional should consider while making a decision and are not necessarily either inclusive or exhaustive. In fact, with identical data in hand, industrial hygienists may come to divergent conclusions. Remember that in this part of the flow diagram the available data have already been proven inadequate for a strongly supported statistical decision. Any HEG ending up in this part of the decision diagram deserves priority and continuing attention from the industrial hygiene surveillance program.

The final two points of emphasis are in the lower right corner of Figure 5.2. First, any evidence of overexposure should trigger corrective measures. Personal protective equipment should become a condition of continuing the offending process, providing enough time for engineering controls to be designed and installed. Second, documentation of results is required not only by law but is part and parcel of good professional practice.

Outcome of Interpretation and Decision Making

The outcome of this part of the exposure assessment strategy is a decision: The exposures are either unacceptable or acceptable. If unacceptable, the next step is to provide employee protection and to institute control measures that will reduce worker exposures. Once controls are in place, a new exposure data set should be sampled, and a new analysis of acceptability of exposures should be performed. In either case, all information leading to the decision should be documented, and recommendations for future sampling needs and exposure control strategies should be made. Documentation is the focus of the next chapter.

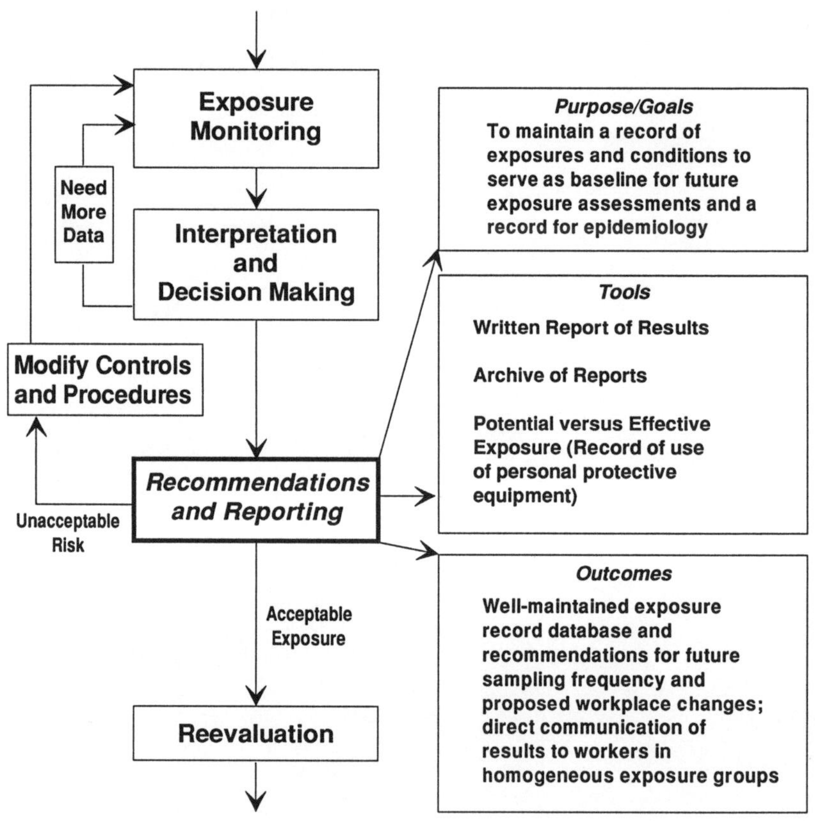

Figure 6.1. Recommendation and reporting component of the exposure assessment strategy

6
Recommendations and Reporting

PURPOSE

Outlined below and in Figure 6.1 are the essential elements of an occupational exposure data repository. Industrial hygienists are encouraged to tailor the suggested structure to match the requirements and culture of their organization.

OVERVIEW OF TOOLS

Each exposure assessment must be documented with a written summary report including all relevant information. This report is the primary vehicle for communicating survey results to workers and also to management. Exposure records are extremely important for a successful, continuing hygiene program. The records serve as a starting point for each new exposure assessment and serve as a baseline to allow detection of upward or downward trends in exposure levels. Documentation is essential for communicating exposure information in response to a worker's complaint or emergency; the records are a quick source of information about previous exposures, process facts, and health-effects information for the environmental agents.

The records are also a vitally important component of future epidemiologic studies. If the records are complete and well-maintained, the homogeneous exposure groups (HEGs) can be easily identified, and estimated exposures can be assigned to workers on the basis of monitoring results in the various groups. The historical data may be helpful to physicians investigating clinical manifestations of disease.

Comprehensive Workplace File

A unique identifier should be assigned to each plant, process area, or workplace. This identification should correspond to a single file that is the permanent repository of information on the workplace of interest. Information in the file should include process information, job and task descriptions, assigned HEGs, environmental agent descriptions (health-effect information and exposure guidelines), monitoring methods and quality assurance procedures, employee complaint investigations, monitoring protocols, raw data, sampling sheets, and reports of exposure assessments.

The process should be completely described in words and by flow diagram. The process area should be shown in sketched floor plans, and points of industrial hygienic interest should be noted (sources, high noise or radiation areas, ventilation hoods, etc.).

In the process description, the process chemistry (including intermediates) should be described, and all environmental agents should be listed. For each agent with high exposure potential, the potential health effects should be fully described. Available standards (permissible exposure limits [PELs], threshold limit values[TLVs®], or internal guidelines) should be referenced and internal occupational exposure limits (OELs) established (concentration and averaging time for each agent).

A record should be made of an industrial hygienist's determination of HEGs. Workers in a plant should be identified by groups, and any further description of performed tasks should be included. If HEGs change from year to year, the HEGs for each year should be retained for future epidemiologic applications.

Any Occupational Safety and Health Administration (OSHA) inspection history should be included as well as internal corporate audits of hygiene programs. Specific employee information such as comments, complaints of odors, and known off-the-job exposures may be useful. Employees concerned about health and safety are often very good plant contacts and are able to identify potential problems.

Any monitoring protocols (plans and strategies for monitoring campaigns) should be included, whether they were for routine monitoring, baseline monitoring, or for specific exposure assessments. These documents should contain the professional rationale for allocating sampling resources.

The last and most important component of the workplace file is a complete set of reports of the exposure assessment performed in the workplace. These reports, which are described in the next section of this chapter in some detail, contain new and analyzed exposure data, professional interpretations, process information, and recommendations for future action.

The Exposure Assessment Report

An industrial hygienist performing an exposure assessment should take the time to document both the work and decision making fully. The report serves as the basis for present and future control actions in the workplace and serves as an information base for the next exposure assessment. As such, it should stand on its own. Timeliness cannot be overemphasized. Memos documenting recommended changes in controls or use of personal protective equipment should be sent immediately. The final report, including these recommendations, should be completed promptly. A delay of even a few weeks can cause loss of memory in the author and loss of confidence from labor and management alike. Report information should be directly conveyed to workers who participated in the industrial hygiene survey.

The following sections are recommended for each exposure assessment report.

1. The (executive) summary: This short (one-page or less) summary should describe the purpose of the exposure assessment, overall results and interpretation, what agents were monitored, the major recommendations growing out of the study, and the recommendations for future monitoring (frequency and substances). This succinct description must be carefully written to provide the relevant information for an overview of the report. Often this sheet will be routed to managers who want the big picture without having to review the often lengthy report that follows.

2. Purpose: The reason for performing the exposure assessment must be clearly stated. Was it a routine, periodic assessment, or was it in response to an employee complaint or process change? This information is very important for later interpretation of exposure data in the context of epidemiology and compliance evaluation.
3. Conclusions: On a point-by-point basis, the major conclusions of the survey should be described. The acceptability of exposures is the primary information described in this section. Also, any diagnostic monitoring and the operational status evaluation of installed exposure control equipment should be reported.
4. Recommendations: In response to the conclusions just described, recommendations for action should be listed. The most important are recommendations for process or control changes and recommendations for changes in use of personal protective equipment. These can be abstracted from action memos sent to management earlier, as soon as the need becomes apparent. Special interest items for future exposure assessments (e.g., a diagnostic survey of an aging lab hood) should be documented. Finally, the frequency of future assessments should be recommended and the agents of interest should be noted. Often, establishing different monitoring frequencies for different agents is reasonable.
5. Exposure evaluation criteria: The relevant exposure guidelines for the environmental agents should be drawn up in tabular format (TLVs, PELs, workplace environmental exposure levels [WEELs], and internal guidelines). The rationale for selecting the guidelines used to set the OELs for this assessment should be discussed. If statistical tests are used later, they should be mentioned here in the context of the OELs employed.
6. Health effects by agent: For those agents monitored, a summary of known health effects (animal or human) should be written (e.g., a paragraph for each agent). The species, the types of effects, and the doses causing the adverse effects should be documented. Any

information about relevant biological rates should be summarized.

7. Monitoring and analytical methods: The monitoring and analytical methods should be briefly described, and references should be given to the full-length procedures used. For each method, the overall measurement error should be given (e.g., coefficient of variation for a method) and the possibility of systematic error explored.
8. Quality assurance: Usage of blanks, spiked samples, or other techniques should be described briefly and results noted.
9. Discussion of results: Monitoring results for each agent in each HEG should be summarized in tabular format. Methods for choosing sampling days and workers should be described (i.e., was selection random?). If statistics were applied in decision making, they should be shown in an appendix. The text in this section should state the conclusion and its basis (statistics, professional judgment, or both). Any diagnostic and area monitoring should be fully described here along with conclusions. Potential exposures (personal protective equipment used) should be clearly distinguished from effective exposures occurring at human body interfaces.
10. Recommendations: All recommendations should be clearly stated. An action officer for each recommendation and a follow-up date should be identified.
11. Job descriptions/homogeneous exposure groups: Job and task descriptions either should be included in the report or should be clearly referenced. HEGs should be described here in some detail.
12. Process description: The process should be described fully in words and in a flow diagram. The chemistry and physics of each task in each process must be complete enough to show locations of potential overexposures. Building diagrams (with points of industrial hygiene concern marked) should be included as figures.

13. References: Any sampling and analytical method used should be cited as well as other supporting materials (e.g., health-effect references, past reports, standards, company personnel manual, etc.).
14. Tables of monitoring results: The raw exposure data should be given with at least the following information: date, times, volume sampled, concentration, and description of the sample (person, tasks performed, special features describing the sample).
15. Appendix: All calculations in support of decisions (e.g., statistical tests) should be included, and every overexposure investigation should be documented.

Data Archiving

In addition to the hard copy file containing the information already described, certain other techniques may be helpful to maximize the utility of exposure assessment records for epidemiologic, health surveillance, and compensation claims. The following provides general guidelines in the computerized maintenance of exposure data.

Information systems should be designed so that pertinent questions can be answered quickly with the information contained in the system. Therefore, one must focus on what questions are likely to be posed and what information is necessary to answer the questions adequately. The following are some of the most frequently asked questions.

- Do the sampling or monitoring data indicate that compliance with regulatory standards has been achieved?
- What substances are associated with a particular job classification, task, or work area, and what is the level of exposure?
- What are the appropriate summary qualitative measures of exposure for homogeneous exposure groups that might be useful in epidemiologic studies?
- What effect has a process change, engineering control, administrative change, etc., had on exposure level?

To answer these and other questions, maintaining some of the data in an electronic database is almost necessary. The types of information to be maintained can be thought of as relating to who is in the workplace, where they work, what substances and exposure levels they encounter, when they were exposed, and how they were exposed.

Employee Characteristics Database The primary concern is the ability to identify employees who worked with given agents. The most useful identifiers are both unique and linkable to other databases. These include the employee's complete name, social security number, date of birth, and work number, if unique. Other attributes can be used (race, sex) but often are already available through the personnel database. The social security number is often the best link between medical, financial, and personnel database systems because it is both unique and widely used.

The Workplace The elements needed are those necessary in defining the HEG. A history of HEG changes (i.e., worker tracking system) is necessary to document changes in exposure potential over a worker's career. Often the information will consist of job classification, physical assignment (often at different levels of detail), task descriptions within a job, and other necessary characteristics.

Stressor Inventory Database Stressors must be linkable to the following:

- Monitoring data
- Homogeneous exposure groups
- Individual workers

A system such as Chemical Abstract Service (CAS) numbers should be used to provide a standard method for identifying a chemical or substance and to account for different naming conventions. A historical perspective should be kept by including effective dates on which each substance is present in the HEG.

Monitoring Database The data elements to be recorded in the monitoring database characterize the source of the monitoring data (i.e., the HEG where the data were collected), along with specific characteristics. The following are among the potential items important to characterize a particular monitoring result.

- Chemical number
- Shift
- Sampling date
- Exposure level
- Units of measurement
- Sample and survey number
- Route of exposure
- Type of sample (personal or area)
- Professional judgment statements
- Frequency of contact
- Report or file number

To answer compliance questions, the monitoring database can be used with the methods recommended in Chapter 5. To answer questions concerning which substances are associated with particular jobs, departments, or HEGs, the substance inventory database can be used. The substance inventory and worker tracking databases can be used in epidemiologic studies that utilize qualitative exposure groups. To answer questions concerning the implementation of controls, process changes, etc., hard copy files may be used in conjunction with electronic databases.

Use in Epidemiology

To consider using quantitative data for subsequent epidemiologic studies, elements of the monitoring database can be used in conjunction with the worker tracking database. There are at least two ways to use quantitative data for epidemiologic purposes. One method particularly useful to chronically acting agents is to rely on mean levels for HEGs. Another method, potentially useful for acutely acting agents, is to rely on area and task measurements and work activity profiles that specify the amount of time a particular employee spends in various areas doing tasks applicable to a certain series of measurements.

OUTCOMES

Ideally, a well-documented report for each exposure assessment will be produced containing the elements described in this chapter and a historical file containing a wide array of relevant information maintained. A record of files should be maintained for easy retrieval of information on specific surveys and processes. In addition to hard copy retention of reports and recommendations, maintaining a computer database of exposure assessment results may sometimes be necessary. Retention times for records vary, but typical retention times range from 25 to 75 years.

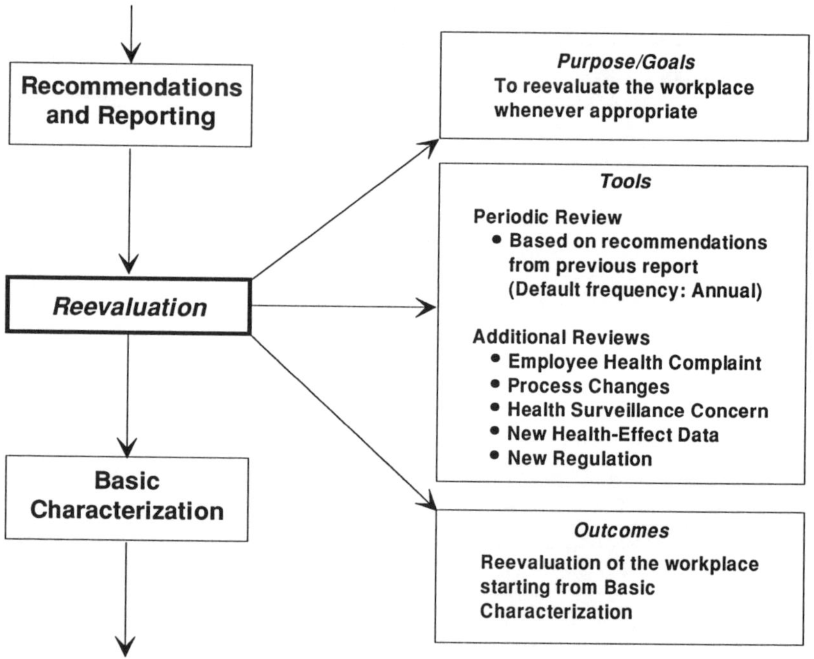

Figure 7.1. Reevaluation component of the exposure assessment strategy

7
Reevaluation

Purpose

Once a workplace has been completely evaluated under the guidelines of the exposure assessment strategy and is considered safe and healthful by the investigating industrial hygienist, the workplace requires less attention. However, every workplace must be reevaluated periodically. This chapter provides guidance for establishing an appropriate reevaluation interval.

Overview of Tools

As stated in Chapter 6, the investigating industrial hygienist's report should recommend a frequency for reevaluation. The typical default frequency is an annual reevaluation of each workplace. Some workplaces may require less frequent reevaluation (e.g., secretarial areas in nonindustrial sites). Others may require frequent attention, such as a monthly review of biological monitoring results (e.g. for organophosphate pesticide production workers).

Periodic Reevaluation

These reevaluations, no matter of what frequency, should be complete, starting with the initial assessment and working through to decisions and reports. Review of the purpose and outcome of each block in the exposure strategy (as shown in Figure 7.1) by the industrial hygienist is important. Detailed monitoring may not be required for each review, but a written plan should document the circumstances under which monitoring would be resumed so the reviewer knows what to look for in the periodic reports.

Additional Reviews

Employee complaint, process change, occupational illness, and other events (such as when new toxicologic data and new regulatory initiatives are revealed) warrant *immediate* reevaluation of the workplace.

Employee Complaint If workers believe that they are suffering adverse health effects from an environmental agent, the workers and their colleagues in the relevant homogeneous exposure group (HEG) should have their exposures evaluated. If the industrial hygienist decides controls are needed, controls should be applied. If not, the industrial hygienist (along with other members of the occupational health team) should meet with the workers to discuss their exposures and their complaints.

Process Change When the industrial hygienist becomes aware of process changes, reevaluation of the workplace is necessary. The HEGs, tasks, environmental agents, and exposure levels may be altered by a process change. If they are, a complete reevaluation is needed. Examples of process changes include a new filtration system in a chemical plant, a new welding system in an automobile plant, a new chemical process in a research and development lab or pilot plant, or an increase in production rate. Plant production and engineering personnel should be reminded to coordinate all such changes with the industrial hygienist prior to implementation.

Occupational Illness Any real or suspected occupational illness detected by a health surveillance program warrants immediate reevaluation of the workplace. Health surveillance is the last line of defense in worker health protection, and any warning signs in this system must be evaluated. A very complete and thorough exposure assessment should be undertaken and coordinated with medical, epidemiologic, and toxicologic efforts. An example of this includes the clinical detection of angiosarcomas and their subsequent linking to vinyl chloride exposures.

Other Events Triggering Reevaluation Additional reevaluations of the workplace would be warranted by factors such as new toxicologic data and new regulatory initiatives.

OUTCOME

In summary, every workplace must undergo periodic reevaluation (annually, unless specified otherwise). Reevaluation is also immediately triggered by employee complaint, process change, health surveillance concerns, new toxicology data, and new regulatory action.

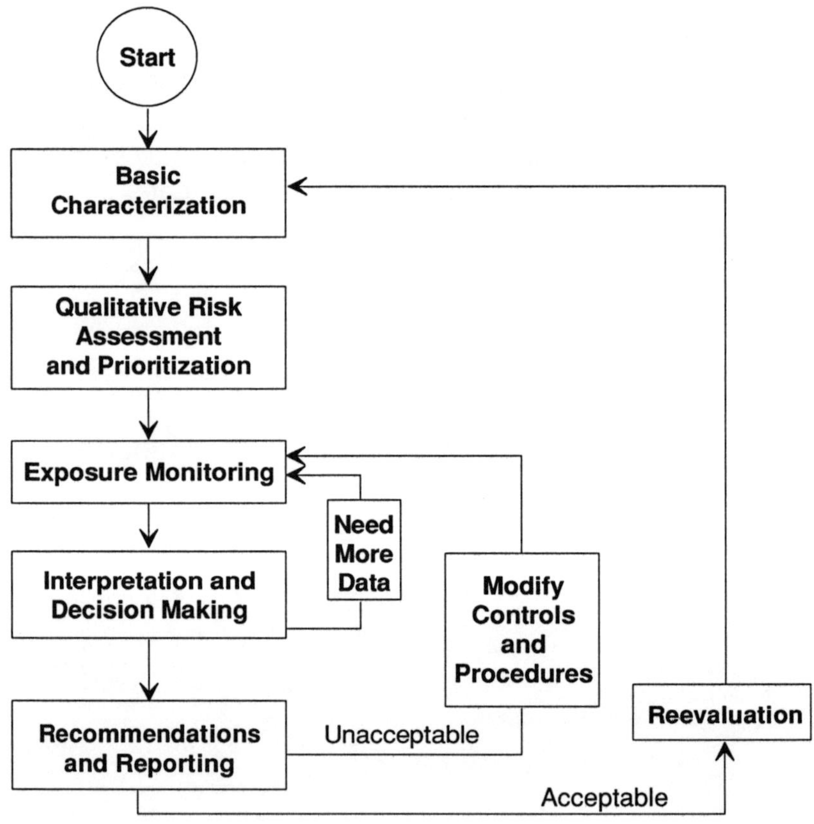

Figure 8.1. Summary of the overall flow diagram of the exposure assessment strategy

8
Conclusion

The overall flow diagram in this document (Figure 8.1) looks and feels familiar because it is simply a formal statement of what industrial hygienists do in everyday practice. Each chapter of this manual has focused on one block in the flow diagram and contains guidance on applying tools in each. Every tool, every guidance offered, has been developed in order to help practicing industrial hygienists achieve one goal: effective allocation of industrial hygiene resources for protecting worker health.

A key message of the document is that there is a great deal more to an occupational exposure assessment strategy than statistical analysis. Statistical analysis is simply an aid to decision making; the primary prerequisite for applying any exposure assessment strategy consistently and accurately is solid professional judgment. This document has attempted to define when and how professional judgment is needed during exposure assessment and has itemized some questions that an industrial hygienist must answer to apply professional judgment consistently.

The occupational exposure assessment strategy is a framework to assist a practicing industrial hygienist. Each block is essential to a full assessment. There is much flexibility in this strategy, and industrial hygienists are encouraged to tailor the strategy to suit the circumstances of their workplace and workers.

Appendix I
An Example of an Exposure Assessment Strategy for a Hypothetical Sodium Chloride Production Plant

OVERVIEW OF THE EXAMPLE

A hypothetical sodium chloride production plant was developed to illustrate the various steps in preparing and implementing an exposure assessment strategy. The strategy starts with the basic characterization of the plant operations and then shows how to apply qualitative risk assessment and prioritization methods; these methods, though different from those presented earlier, are developed to emphasize that any procedure can be used as long as all relevant factors (i.e., exposures, health effects, homogeneous exposure groups [HEGs], task analysis, etc.) have been taken into account.

For the monitoring portion of the strategy, an exposure model was developed to indicate what levels of exposure might be expected in a monitoring survey. This approach is useful in deciding which of the available monitoring techniques would be most effective in providing the necessary data. The priority ratings and the exposure model were used to establish an initial monitoring plan for those job classifications with the highest potential for exposure.

For the interpretation and decision-making step of the assessment strategy, hypothetical monitoring data were developed for one of the job classifications. The data were evaluated by percent exceedence and analysis of the mean techniques to determine the acceptability of the exposure levels. The statistical tools were used as the basis for recommendations for controls and an ongoing monitoring program.

Sodium Chloride Production Plant
Process Description

Chlorine gas is received directly by pipeline from an adjacent vendor plant; the gas arrives at a line pressure of 200 psig, passes through a pressure let-down valve, and enters T-1 column at 20 psig. 25% hydrogen peroxide is received in tank cars and is diluted with process water to 19% before being pumped to T-2 column. In Columns T-1 and T-2, chlorine and hydrogen peroxide go through a counterflow reaction to produce a 29% aqueous hydrochloric acid (HCl) solution; trace amounts of chlorine are vented to the atmosphere at T-2 column.

A 50% sodium hydroxide solution is received by pipeline from an adjacent vendor plant and is stored in S-2 tank before being pumped to R-1 reactor. From T-2 column, the HCl solution is pumped to R-1 reactor to react with the caustic and produce a 35% sodium chloride solution (brine). The brine is contacted with heated air in F-1 fluid bed dryer to make sodium chloride granules with a size range of 100 to 200 microns. The finished product is transferred to storage hopper H-1 via an airvey system.

The industrial grade sodium chloride is packaged in 50 pound bags and palletized in 1000 pound loads. Each pallet is film-wrapped in an automatic shrink-film apparatus. A tow motor is used to move pallets from the bagging station to the shrink-film station and then to the warehouse. At the warehouse, the tow motor is also used to load the wrapped pallets into truck trailers.

The plant is currently producing about 50 million pounds per year of industrial grade sodium chloride; the plant operates about 300 days per year with the remaining time used for scheduled maintenance work. On a daily basis, the plant receives about 53 tons of chlorine gas, 120 tons of caustic solution, and 306 tons of hydrogen peroxide solution.

Figure I.1. Process description developed for an example of an exposure assessment strategy for a hypothetical sodium chloride production plant. Development of such a workplace description is fundamental to completing the workplace characterization portion of the basic characterization step of the strategy.

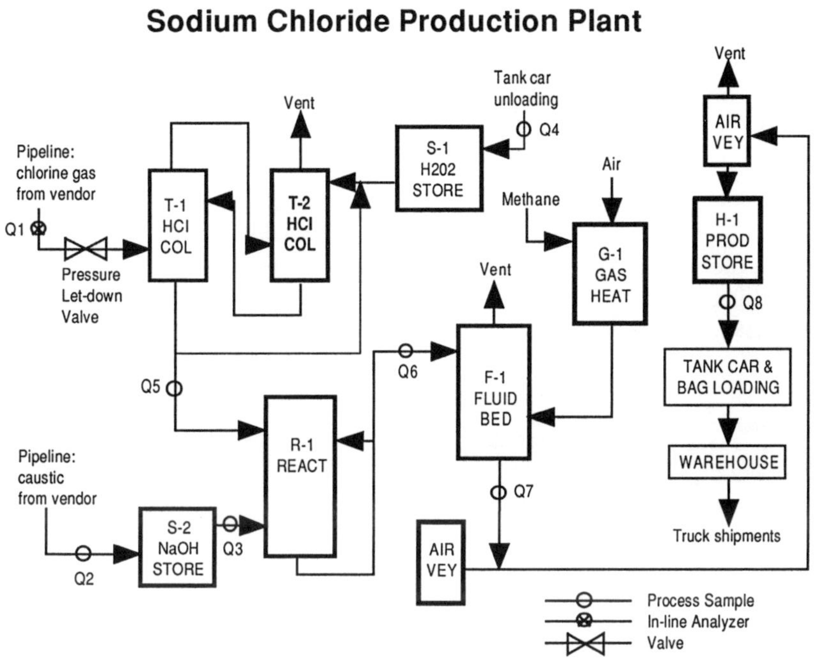

Sodium Chloride Production Plant

Figure I.2. Schematic flow diagram developed as part of the workplace characterization portion of the basic characterization step of the strategy.

BASIC CHARACTERIZATION

To describe the sodium chloride production plant operations, both a process description (Figure I.1) and a schematic flow diagram (Figure I.2) were prepared. The process description outlines the process chemistry and indicates the concentration of the various components. Information is also given about product characteristics as well as product packaging and handling systems; experienced industrial hygienists will recognize the possible noise and carbon monoxide problems associated with tow motor operations. The process description also includes a brief discussion of the production rate and the amount of material handled in the plant.

The flow diagram shows the progression of process materials, from raw materials to finished product, through each of the major unit operations; each operation is labeled by function (reactor, storage tank, etc.). By also including process equipment labels (R-1, T-2, etc.) on each of the unit operations, the

Sodium Chloride Production Plant

Job Descriptions
Administrative and Office Personnel

Individuals in this group spend essentially all of their time on administrative, supervisory, consulting, computer, and/or other office activities. They have little potential for exposure to any of the agents encountered at the sodium chloride production plant. Industrial hygiene inventories are not considered necessary for these job classifications.

Personnel Roster

Name	Company ID No.	Job Classification	Job Code
	30045	Plant Manager	0020
	00073	Sales Representative	0033
	10028	Purchasing Agent	0037
	10033	Secretary	0043
	10055	Shipping & Receiving Clerk	0048

Figure I.3. Sample job description for administrative and office personnel in a hypothetical sodium chloride production plant. The description was developed to fulfill the work force characterization requirements of the basic characterization step of the strategy and to be used as the basis for determining homogeneous exposure groups.

flow diagram can be a useful tool for helping the industrial hygienist to understand the activities of the operating personnel during a survey. The relatively simple sketch of the operations can provide the experienced industrial hygienist with a wealth of information regarding (1) protective equipment requirements at the various process sample points, (2) possible noise problems at the fluid bed and airvey systems, and (3) potential dust problems at the bagging station.

Figures I.3 and I.4 give brief job descriptions for all personnel in the plant. In Figure I.3, administrative and office personnel are combined into an HEG with negligible potential for exposure; this group is separated from the rest because there would be no need to prepare a more detailed qualitative risk assessment for these individuals. Figure I.4 shows the homogeneous exposure groups for the remaining individuals who spend time in the process areas and, therefore, have some potential for exposure. The job descriptions are very brief, giving just enough information to indicate how much time is spent in each of the various process areas.

Also as part of the basic characterization step, a chemical, biological, and physical agent list would be prepared for each of the HEGs; the toxicity of each agent would be evaluated so that an appropriate occupational exposure limit (OEL) could be determined (if one did not exist). With this information in hand, the industrial hygienist would be ready to advance to the next step of the exposure assessment strategy.

Qualitative Risk Assessment/Exposure Ranking

The second step in the strategy requires the industrial hygienist to do a qualitative risk assessment for each HEG/agent combination.

Figure I.5 gives one ordinal system for ranking potential for exposure to a given agent within an HEG. Using the stated definitions for each category, an industrial hygienist should be able to rank most chemicals in an exposure category after reviewing past data, model predictions, and other factors.

Figure I.6 shows a task analysis to confirm estimates of the degree of exposure for a given HEG/agent combination. The system shown in Figure I.5 is used to determine the potential for exposure for each task where the agent in question is encountered. A time-weighting approach is used in the task analysis to estimate the overall potential for exposure as shown for the hydrochloric acid and noise examples. The degree of exposure is assessed for those activities where the agent is encountered; activities with a negligible potential for exposure are not included in the time-weighting process.

After a degree of exposure has been estimated the ordinal system shown in Figure I.7 can be used to establish a priority ranking. With knowledge of the toxicity and potential for exposure, the industrial hygienist must use judgment to determine the risk of overexposure to a particular agent for a given HEG. If a ranking is difficult to establish, an estimate can be made by using a numerical system such as that shown in Figure 3.2 in the main text.

Sodium Chloride Production Plant
Job Descriptions
Operations Personnel

Superintendent: Spends about 10% of time in general process areas observing operations, checking equipment conditions, and supervising maintenance work; remaining time spent in office environments on administrative, supervisory, and planning activities.

Engineer: Spends about 35% of time in general process areas troubleshooting process problems, supervising maintenance work, and collecting industrial hygiene samples; remaining time spent on training, computer program development, and other office activities.

Shift Supervisor: Spends about 5% of time in general process areas checking on operations and investigating possible process problems; remaining time spent in control room areas overseeing operation of the acid, reactor, and dryer systems.

Relief Operator: Spends about 20% of time covering each of the shift supervisor, acid system operator, reactor system operator, and assistant operator job classifications; remaining time spent on various maintenance activities.

Operator, Acid System: Spends about 40% of time in the HCl production areas checking equipment, adjusting flows, and preparing equipment for maintenance; about 10% of time is spent collecting process samples and another 25% is spent in lab running analyses on all process samples; remaining time is spent in control room areas.

Operator, Reactor System: Spends about 60% of time in the reactor and fluid bed dryer areas checking equipment, adjusting flows, and preparing equipment for maintenance. About 10% of time is spent collecting process samples; remaining time is spent in control room areas.

Assistant Operator: Spends about 50% of time at bagging station loading product into 50 pound bags; another 25% of time is spent using tow motor to move pallets to and from the warehouse and to load product into truck trailers; about 5% of time is spent loading product directly from storage into hopper cars; remaining time is spent in control room areas.

Electrical/Instrument Technician: Spends about 5% of time in the HCl production areas maintaining and calibrating in-line chlorine analyzer; approximately 25% of time is spent in general process areas and switchgear room maintaining electrical equipment; about 50% of time is spent in control room or maintenance shop repairing or modifying process control instruments; remaining time is spent on office activities.

Pipe fitter/Welder: Spends about 30% of time in maintenance shop on welding work associated with process piping and equipment repairs; about 40% of time is spent in general process areas removing or replacing piping and process equipment; remaining time is spent in shop office or control room.

Utility Mechanic: Spends approximately 20% of time helping process operators with preparation of equipment for maintenance; about 50% of time is spent repairing equipment in shop; another 10% of time is spent in general process areas making minor repairs to process equipment; remaining time is spent in shop office or in control room.

Personnel Roster

Name	Company ID No.	Job Classification	Job Code
J. R. Smith	20066	Superintendent	0062
M. C. Jones	10081	Engineer	0085
	10018	Shift Supervisor	0105
	10025	Shift Supervisor	0105
	10038	Shift Supervisor	0105
	10040	Shift Supervisor	0105
	10019	Relief Operator	0121
	10050	Relief Operator	0121
	10029	Operator, Acid System	0123A
	10042	Operator, Acid System	0123A
	10066	Operator, Acid System	0123A
	10071	Operator, Acid System	0123A
	10021	Operator, Reactor	0123R
	10048	Operator, Reactor	0123R
	10052	Operator, Reactor	0123R
	10065	Operator, Reactor	0123R
	10088	Assistant Operator	0129
	10096	Assistant Operator	0129
	10099	Assistant Operator	0129
	10100	Assistant Operator	0129
	10044	Elec/Instr Technician	0210
	10030	Pipe fitter/Welder	0230
	10061	Utility Mechanic	0255
	10078	Utility Mechanic	0255

Figure I.4. Sample job description for operations personnel in a hypothetical sodium chloride production plant. The descriptions were developed to fulfill the work force characterization requirements of the basic characterization step of the strategy and to be used as an approach for determining homogeneous exposure groups.

Sodium Chloride Production Plant

Degree of Exposure

N = Negligible; no contact; does not work in areas of the plant where the material is handled (less than 10% of the occupational exposure limit)

L = Low; minor contact; infrequent handling or works in areas of the plant where the material is contained in closed systems (10–50% of the occupational exposure limit)

M = Moderate; daily contact; handles material regularly under the following conditions:

A. No ventilation: material has low vapor pressure or does not generate much dust.

B. Ventilation: material has elevated vapor pressure or is relatively dusty.

(50–100% of the occupational exposure limit)

H = High; gross contact; handles material of high vapor pressure or dustiness on a regular basis (greater than 100% of the occupational exposure limit)

Figure I.5. Sample exposure rating system for estimating exposure to a given agent within an HEG. Note: If monitoring data are available, average measured exposures may be compared to the criteria shown in parentheses to define the degree of exposure for a job classification.

Figures I.8, I.9, and I.10 show the qualitative risk assessment and prioritization summaries for three of the HEGs (operator, acid system; operator, reactor; and assistant operator) in the sodium chloride production plant. In addition to the degree of exposure and priority ranking, the OEL used in the evaluation is included; this same outline can also be used to indicate the type and effective date of the available health hazard data sheet (i.e., material safety data sheet [MSDS]). Toxic effect rating is not given in these tables since a direct judgment was used to establish priority ranking from the OEL and the degree of exposure.

Selecting agents with priority ratings of 1 to 3 for highest priority for monitoring, the industrial hygienist can now establish monitoring requirements for each HEG; noise would be the agent of highest priority since two of the HEGs had a priority rating of 2. Chlorine, sodium hydroxide, hydrochloric acid, and carbon monoxide each had a priority rating of 3; the industrial hygienist must use judgment and knowledge of the process (odor level, complaints of irritation, etc.) to decide, first, if monitoring is really necessary and, second, the order of importance if monitoring is to be conducted.

MONITORING PLAN

Once the agents to be monitored have been determined, a plan is needed to outline the types of samples that will be collected; the plan would be based on the availability of survey equipment, the priority ranking of HEGs, and the status of work being done at the workplace. For the purposes of this example, the assumption was made that the appropriate monitoring equipment would be available to collect the kinds of samples needed to characterize the employee exposures.

In the sodium chloride production plant, the chronic effects of overexposure to noise would be one concern. Therefore, the survey work should aim at defining the long-term average exposures based on 8-hr exposure measurements. However, for control purposes, an area survey would be necessary to associate noise levels with specific process equipment (diagnostic sampling survey).

Chlorine, sodium hydroxide, and hydrochloric acid are strong irritants; therefore, the acute effects of overexposure would be of primary interest. The monitoring plan should be designed to determine peak airborne concentrations during specific tasks with the highest potential for exposure to compare against the ceiling limits. Using time-weighted average (TWA) techniques an industrial hygienist can use the task-oriented results to estimate average 8-hr exposures. For chlorine, a long-term average OEL that factors in acceptable short-term excursions was to be used for evaluating acceptability of chlorine exposures. For hydrochloric acid, a short-term excursion OEL was used; 15-min TWAs during tasks were monitored.

Carbon monoxide levels that might be expected during tow motor operations make the acute effects of carbon monoxide

Sodium Chloride Production Plant
Estimating Degree of Exposure by Task Analysis for the Operator, Acid System

Summary

Area/Task	Total Shift (%)	Frequency of Task per Shift	Task Duration (minutes)	Degree of Exposure for Each Agent[A]					
				HCl	Cl_2	H_2O_2	NaOH	Acetone	Noise
Control Room	25	1	120	1	1	1	1	1	1
Laboratory	25	2	60	2	1	2	2	3	2
Production Areas									
Sampling	10	2	24	4	2	2	1	1	3
Equip Check	20	2	48	2	2	2	1	1	3
Adjust Flows	10	2	24	2	2	2	1	1	3
Prep for Maint	10	1	48	3	3	3	1	1	3

Examples

Example 1: Hydrochloric Acid

Area Task	Duration (minutes)	Degree of Exposure	Weighted Ratio[B]
Control Room		1	0.00
Laboratory[C]	20	2	0.38
Production Areas			
Sampling[C]	20	4	0.77
Equip Check[C]	32	2	0.62
Adjust Flows[C]	16	2	0.31
Prep for Maint[C]	16	3	0.46
Totals	104		2.54
Est. TWA[D] Degree of Exposure		3	

Example 2: Noise

Area Task	Duration (minutes)	Degree of Exposure	Weighted Ratio[B]
Control Room		1	0.00
Laboratory	120	2	0.67
Production Areas			
Sampling	48	3	0.40
Equip Check	96	3	0.80
Adjust Flows	48	3	0.40
Prep for Maint	48	3	0.40
Totals	360		2.67
Est. TWA Degree of Exposure		3	

[A]Values for degree of exposure; N = 1, L = 2, M = 3, H = 4. (Figure I.5 defines N, L, M, and H.)

[B]Weighted Ratio = (Duration) (Degree of Exposure)/(Total Encounter Time).

[C]Assumption: Total Task Time is allocated evenly for each chemical agent encountered during the activity.

[D]TWA = Time-weighted average.

Figure I.6. Sample task analysis used for estimating the degree of exposure experienced by the operator, acid system, of the hypothetical sodium chloride production plant

Sodium Chloride Production Plant

Priority Ranking

1. High risk: Requires immediate action to reduce the level of exposure and follow-up monitoring to determine the effectiveness of the control measures
2. Elevated risk: Requires a monitoring program to determine the level of exposure so that the appropriate controls can be implemented; also requires regular follow-up monitoring to make sure that the protective devices are providing adequate control
3. Moderate risk: Requires overt action in the form of protective devices, special training programs, and occasional monitoring to ensure that exposures are maintained under control
4. Low risk: Requires educational review of the consequences of overexposure and the proper precautions in case of a nonroutine event
5. Minimal risk: Requires availability of health hazard information in the plant's industrial hygiene manual. Low toxicity or very low potential for exposure indicates that no further effort is required.

Figure I.7. Sample priority ranking system for evaluating exposures in a hypothetical sodium chloride production plant

exposure a concern, but only for prolonged time periods (i.e., several hours). Exposure levels could adequately be characterized by 8-hr measurements combined with several short-term excursion measurements during periods of heavy tow motor usage.

Using priority rankings based on the review of the pertinent toxic effects and potential exposure levels, an initial monitoring plan (Figure I.11) was developed for each HEG to be included in the surveillance program. The plan includes the type and number of samples to be collected on a random basis during the current year. However, the plan should remain flexible and should be modified if unexpectedly high exposure levels are found during the surveillance work.

Sodium Chloride Production Plant
Industrial Hygiene Inventory for March 1989

Process (0001):	Sodium Chloride Production
Manager/IH Contact:	J. Smith/M. Jones
Industrial Hygienist:	G. Flores
Job Class (0123A):	Operator, Acid System

Chemical/Physical Agents	OEL	Degree Exposure	Priority Rating
Raw Materials			
Chlorine	0.5 ppm	L	3
Hydrogen Peroxide	1.5 mg/m^3	L	3
Sodium Hydroxide	2 mg/m^3 C[A]	L	4
Intermediates			
Hydrochloric Acid	5 ppm C	M	3
Brine Solution	10 mg/m^3	M	5
By-Products			
Oxygen		L	5
Aux. Proc. Matl.			
LPG	1000 ppm	N	5
Methane	Asphyx.	N	5
Polyethylene	10 mg/m^3	N	5
Lab Chemicals			
Acetone	750 ppm	M	4
Potass. Hydroxide	2 mg/m^3 C	L	4
Hydrochloric Acid	5 ppm C	L	4
Phenolphthalein	Not Est.	L	4
Products			
Sodium Chloride	10 mg/m^3	L	5
Miscellaneous			
Carbon Monoxide	50 ppm	N	4
Phys. Stresses			
Noise	85 dBA	M	3

[A]C = Ceiling.

Figure I.8. Sample qualitative risk assessment and prioritization summary for the operator, acid system, HEG in a hypothetical sodium chloride production plant. OEL = occupational exposure limit.

Sodium Chloride Production Plant
Industrial Hygiene Inventory for March 1989

Process (0001): Sodium Chloride Production
Manager/IH Contact: J. Smith/M. Jones
Industrial Hygienist: G. Flores
Job Class (0123R): Operator, Reactor

Chemical/Physical Agents	OEL	Degree Exposure	Priority Rating
Raw Materials			
Chlorine	0.5 ppm	N	4
Hydrogen Peroxide	1.5 mg/m^3	N	4
Sodium Hydroxide	2 mg/m^3 C[A]	N	3
Intermediates			
Hydrochloric Acid	5 ppm C	L	4
Brine Solution	10 mg/m^3	L	5
By-Products			
Oxygen		N	5
Aux. Proc. Matl.			
LPG	1000 ppm	N	5
Methane	Asphyx.	L	4
Polyethylene	10 mg/m^3	N	5
Lab Chemicals			
Acetone	750 ppm	L	4
Potass. Hydroxide	2 mg/m^3	N	4
Hydrochloric Acid	5 ppm C	N	4
Phenolphthalein	Not Est.	N	5
Products			
Sodium Chloride	10 mg/m^3	M	5
Miscellaneous			
Carbon Monoxide	50 ppm	L	4
Phys. Stresses			
Noise	85 dBA	M	2

[A]C = Ceiling.

Figure I.9. Sample qualitative risk assessment and prioritization summary for the operator, reactor, HEG in a hypothetical sodium chloride production plant.

Sodium Chloride Production Plant
Industrial Hygiene Inventory for March 1989

Process (0001): Sodium Chloride Production
Manager/IH Contact: J. Smith/M. Jones
Industrial Hygienist: G. Flores
Job Class (0129): Assistant Operator

Chemical/Physical Agents	OEL	Degree Exposure	Priority Rating
Raw Materials			
Chlorine	0.5 ppm	N	4
Hydrogen Peroxide	1.5 mg/m^3	N	4
Sodium Hydroxide	2 mg/m^3 C[A]	N	4
Intermediates			
Hydrochloric Acid	5 ppm C	N	4
Brine Solution	10 mg/m^3	N	5
By-Products			
Oxygen		N	5
Aux. Proc. Matl.			
LPG	1000 ppm	M	4
Methane	Asphyx.	N	5
Polyethylene	10 mg/m^3	L	5
Lab Chemicals			
Acetone	750 ppm	N	5
Potass. Hydroxide	2 mg/m^3	N	4
Hydrochloric Acid	5 ppm C	N	4
Phenolphthalein	Not Est.	N	5
Products			
Sodium Chloride	10 mg/m^3	M	5
Miscellaneous			
Carbon Monoxide	50 ppm	M	3
Phys. Stresses			
Noise	85 dBA	M	2

[A]C = Ceiling.

Figure I.10. Sample qualitative risk assessment and prioritization summary for the assistant operator in a hypothetical sodium chloride production plant.

Sodium Chloride Production Plant

Initial Monitoring Plan

The qualitative risk assessment and priority setting results indicate that process operators and maintenance personnel appear to have a significant potential for exposure to noise, chlorine, hydrochloric acid, sodium hydroxide, and hydrogen peroxide. Applying the knowledge of the process that was gained through the various basic characterizations, task analysis, and exposure modeling techniques, an industrial hygienist developed the following monitoring plan for the current year.

	Activities to be Monitored		
Job Classification	Process Sampling (samples/duration)	Preparing Equipment for Maintenance (samples/duration)	Full-Shift Activities (samples/duration)
Acid System Operator			
Noise[A]	[B]	[B]	6/8-hr TWA
Chlorine	10/15-min STEL	10/15-min STEL	10/8-hr TWA
Hydrochloric Acid	10/15-min Peak	10/15-min Peak	[B]
Hydrogen Peroxide	[B]	[B]	10/8-hr TWA
Reactor Operator			
Noise[A]	[B]	[B]	6/8-hr TWA
Sodium Hydroxide	10/15-min STEL	10/15-min STEL	6/8-hr TWA

Packaging Operator			
Noise[A]	[B]	[B]	6/8-hr TWA
Carbon Monoxide[C]	[B]	[B]	10/8-hr TWA
Utility Mechanic			
Chlorine	[B]	10/15-min STEL	[B]
Hydrochloric Acid	[B]	10/15-min Peak	[B]
Sodium Hydroxide	[B]	10/15-min STEL	[B]

[A]Noise dosimetry will be supplemented with a thorough area noise survey to define high noise sources.

[B]No samples planned.

[C]Carbon monoxide dosimetry will be done with a direct-reading instrument and data logger; so 15-min excursions will also be evaluated.

Figure I.11. Sample initial monitoring plan developed as part of the exposure assessment strategy for a hypothetical sodium chloride production plant

Interpretation and Decision Making

After the monitoring samples have been collected, the industrial hygienist must arrive at some conclusions about the acceptability of the exposure levels. Even if all of the measurements are within the guidelines, there is still a need to check the possibility that some exposures might exceed the established limits. This consideration is particularly important for agents with short-term or ceiling limits established to avoid acutely toxic effects.

For the acid system operators, hypothetical data were generated for each of the agents included in the monitoring plan. Arithmetic and geometric means and standard deviations were calculated for each set of data. For measurements compared to an 8-hr limit, the confidence limits on the arithmetic mean were calculated. For measurements compared to a short-term or ceiling limit, a percent exceedence test was performed using one-sided tolerance limits.

Figure I.12 outlines the evaluation of six 8-hr measurements of noise exposure ranging from 79 dBA up to 92 dBA, using 85 dBA as the acceptable level of exposure for an 8-hr period. The dBA values were converted to a percent of the allowable limit by dividing the exposure time (480 min) by the allowed time (480 min for 85 dBA, 960 min for 80 dBA, etc.).

Figure I.13 shows the evaluation of 10 measurements (8-hr average) of chlorine exposure ranging from nondetectable (<0.1 ppm) up to the current occupational exposure limit of 0.5 ppm.

Figure I.14 gives the results of the evaluation of 10 excursion measurements (15-min average) of hydrochloric acid exposure ranging from 0.2 to 4.0 ppm, which were compared to the ceiling limit of 5 ppm.

Using these statistical tools, the industrial hygienist is in a better position to define the overall range of possible exposure levels. This, in turn, should improve the reliability of recommendations to ensure that exposures are maintained within acceptable levels.

Sodium Chloride Production Plant
Evaluation of Noise Exposures

Job: Acid operator
Occupational exposure limit: 85 dBA, 8-hr average (Assumed LTA)

Random full-shift, 8-hr TWA results (dBA):
80 79 80 92 81 88

Report descriptive statistics:

Convert dBA values to % PEL:

dBA	% PEL
80	50
79	44
80	50
92	264
81	57
88	152

n = 6
Range: 79–92 dBA
% n > OEL: 34

Arithmetic mean: 103% (85 dBA)
Arithmetic standard deviation: 88.9%
Geometric mean: 80% (83 dBA)
Geometric standard deviation: 2.10

Evaluate confidence limit on mean (Appendix III)

$\bar{\chi} = 103\%$

$$\text{UCL} = 103\% + \frac{2.571\ (88.9)}{\sqrt{6}}$$

$$\text{LCL} = 103\% - \frac{2.571\ (88.9)}{\sqrt{6}}$$

[10%, 196%]

Investigate the two exposures greater than the OEL. There are insufficient data to make a decision about the mean noise exposure since the confidence interval contains 100%. Either take more samples or evaluate on basis of professional judgment.

Figure I.12. Sample statistical interpretation and evaluation of noise exposures at a hypothetical sodium chloride production plant

Sodium Chloride Production Plant
Evaluation of Chlorine Exposures

Job: Acid operator
Occupational exposure limit: 0.5 ppm 8-hr TWA
(Assumed LTA)

Random full-shift, 8-hr TWA results (ppm):
<0.1 0.3 0.4 0.1 <0.1 0.5 0.2 <0.1 0.2 0.3

Report descriptive statistics
(nondetected exposures were calculated at 1/2 LOD):

n = 10
Range: <0.1–0.5 ppm
% n > OEL = 0

Arithmetic mean: 0.22 ppm
Arithmetic standard deviation: 0.16
Geometric mean: 0.16
Geometric standard deviation: 2.5

Evaluate confidence limit on mean (Appendix III)

Assuming lognormality of data:

$\bar{\chi}_{\ln x} = -1.83 \qquad S_{\ln x} = 0.92 \qquad t_{\frac{\alpha}{2}, \upsilon=9} = 2.262$

$$\bar{\chi} = e^{-1.83} e^{\frac{(0.92)^2}{2}} = 0.25 \text{ ppm}$$

$$\text{LCL} = e^{\left[-1.83 - 2.262 \frac{0.92}{\sqrt{10}}\right]} e^{\frac{(0.92)^2}{2}} = (0.08)(1.53) = 0.13 \text{ ppm}$$

$$\text{UCL} = e^{\left[-1.83 + 2.262 \frac{0.92}{\sqrt{10}}\right]} e^{\frac{(0.92)^2}{2}} = (0.31)(1.53) = 0.47 \text{ ppm}$$

95% CI = [0.13, 0.47] Confidence interval does not contain the OEL and is less than OEL; therefore, mean exposure is acceptable.

Figure I.13. Sample statistical interpretation and evaluation of chlorine exposures at a hypothetical sodium chloride production plant

Sodium Chloride Production Plant
Evaluation of Hydrochloric Acid Exposures

Job: Acid operator
Task: Process Sample Collection
Occupational exposure limit: Ceiling 5 ppm (Excursion Limit)

Random peak exposure samples (ppm):

0.3 2.5 3.0 0.2 2.3 2.0 4.0 3.2 1.9 3.4

Report descriptive statistics:

n = 10
Range: 0.2–4.0 ppm
% n > OEL: 0

Arithmetic mean: 2.28 ppm
Arithmetic standard deviation: 1.25
Geometric mean: 1.67 ppm
Geometric standard deviation: 2.84

Evaluate % exceedence by one-sided lognormal tolerance limit:

$\bar{\chi}_{lnx} = 0.51$

$S_{lns} = 1.04$

$n = 10$

K (95%) confidence that no more than 5% exceed the value) = 2.911

$UTL = e^{(\bar{\chi}_{lnx} + K(S_{lnx}))} = e^{(0.51) + 2.911(1.04)}$

$= 34.4$ ppm

No conclusive statement is possible based on the tolerance limit approach. (One can have 95% confidence that no more than 5% of the exposures are greater than 34 ppm.) With a 5 ppm ceiling, this does not reveal much. Tolerance limit approaches require large sample sizes to make a definitive statement. However, since all 10 exposures are less than the OEL, a decision of "acceptability" can be made on the basis of professional judgment (or more samples can be collected).

Figure I.14. Sample statistical interpretation and evaluation of hydrochloric acid exposures at a hypothetical sodium chloride production plant

REEVALUATION FREQUENCY

As part of the documentation process for the exposure assessment, recommendations for future reevaluation should be made. For example:

Groups not monitored:	Biannual reevaluation (every 2 years)
Acid Operator:	
Noise:	Reevaluate noise exposures within 6 months since two measured exposures were over limit. Diagnose cause of overexposure using area and task monitoring.
Chlorine:	Reevaluate biannually.
Hydrochloric Acid:	Reevaluate annually since some exposures approached the OEL.

Appendix II
Descriptive Statistics and Probability Plotting

Overview

Many industrial hygiene data sets can be interpreted by comparing the occupational exposure limit (OEL) to descriptive statistics or to a graphical plot of the data. When most of the data are clustered well below or well above the OEL, a decision on workplace acceptability can generally be made by using descriptive statistics and professional judgment. When the range of data approaches or includes the OEL, inferential statistics become useful for decision making (Appendixes III, IV, and V). In these cases, the descriptive statistics are used to support the professional judgment underlying the suitability of the inferential statistics used. Assumptions like normality or lognormality of data can be easily checked with descriptive statistics.

Appendix II is intended to provide a refresher for descriptive statistics and, as such, will not replace a standard statistics textbook for beginners. Statistical approaches useful in some situations and not addressed in the appendixes include non-parametric inferential statistics, binomial statistics, and extreme value statistics. The industrial hygienist is responsible for verifying the statistical assumptions underlying data interpretation tools. Retaining a professional statistician to select valid approaches that apply maximum available statistical power is often advisable.

Appendix III contains instructions for testing the confidence interval on the mean of data, and Appendix IV contains instructions for testing the tolerance interval for both normal and lognormal data sets. None of these tests should be trusted unless the industrial hygienist using them is confident that the data being analyzed are adequately described as either normally or lognormally distributed. Appendixes II, III, and IV are designed to be read as a unit.

Descriptive Statistics

Any set of data may be analyzed to calculate a measure of central tendency and a measure of spread or variability. Measures of central tendency that are obvious by inspection are the mode and the median. The mode is the value that occurs most frequently. The median is the value in the middle—half the data lie above it and half lie below it. A measure of spread that is obvious by inspection is the range, the difference between the maximum and the minimum values in the data.

For inferential analysis, mathematically "nice" parameters are often used to describe appropriate data sets. For well-behaved data sets, as few as two parameters may completely describe the distribution of data. If the data are normally distributed, the mean and standard deviation completely describe the probability density function. If the data are lognormally distributed, the geometric mean and geometric standard deviation completely describe the probability density function.

A common statistical convention uses (as indicated below) uppercase letters to describe parameters estimated from data, lowercase letters to describe individual values of the data points, and Greek letters to represent the true (but often unknowable) parameters of the population being sampled.

x	=	value of a data point
R	=	range of data, (maximum–minimum)
M	=	(sometimes $\bar{x}$) arithmetic mean of data
μ	=	true population mean (Greek letter mu)
S	=	sample standard deviation, computed from data
σ	=	true population standard deviation (Greek letter sigma)
GM	=	geometric mean of data
gμ	=	true geometric mean of population

GS	=	geometric standard deviation of data
gσ	=	true geometric standard deviation of population
γ	=	confidence, or probability that a hypothesis is true
n	=	number of measurements in a sample of the population
υ	=	statistical degrees of freedom (n–1)
P	=	Proportion of population in hypothesis (i.e., $P \leq UTL$)
UCL	=	upper confidence limit on μ, the estimated mean
LCL	=	lower confidence limit on μ, the estimated mean
UTL	=	upper tolerance limit on data (normal distribution)
LTL	=	lower tolerance limit on data (normal distribution)
GUCL	=	upper confidence limit on gμ (lognormal distribution)
GLCL	=	lower confidence limit on gμ (lognormal distribution)
GUTL	=	upper tolerance limit on data (lognormal distribution)
GLTL	=	lower tolerance limit on data (lognormal distribution)
t_p	=	upper bound of 100P% of the Student *t* distribution
z_p	=	$(x_p - \mu)/\sigma$, upper bound of 100P% of normal distribution

The parameters M, S, GM, and GS can be computed for any set of data by applying the simple equations below. Professional judgment is required to decide whether these parameters represent the data well enough to be used in inferential statistical tests involving confidence limits or confidence intervals. The rest of Appendix II is designed to illustrate the features of data sets that make them suitable for inferential analysis by lognormal or by normal statistics.

Four equations can be used to calculate the four main parameters. The mean, M, and standard deviation, S, are easily computed on most scientific calculators and computers by using Equations II.1 and II.2.

$$M = \frac{\sum_{i=1}^{n} (x_i)}{n} \quad \textbf{(II.1)}$$

$$S = \frac{\sqrt{\sum_{i=1}^{n} [x_i - M]^2}}{n-1} \quad \textbf{(II.2)}$$

The geometric mean, GM, and the geometric standard deviation, GS, are easily computed by the same calculators by using Equations II.3 and II.4. First, one must create a log-transformed data set by taking the natural logarithm of each value in the data set. Then, one must use the log transformed data as input to the standard scientific calculator. The calculator key labeled *mean* returns ln(GM), and the key labeled *standard deviation* returns ln(GS). Using the exponential or antilog function to calculate the value of the GM and the GS, then, is simple.

$$\ln(GM) = \left(\frac{\sum_{i=1}^{n} \ln(x_i)}{n} \right) \quad \textbf{(II.3)}$$

$$\ln(GS) = \sqrt{\frac{\sum_{i=1}^{n} \left[\ln(x_i) - \ln(GM)\right]^2}{n-1}} \quad \textbf{(II.4)}$$

A dust that has an OEL of 10 mg/m^3 can be used to illustrate these procedures. Full-shift total dust samples were randomly collected from workers in a homogeneous exposure group (HEG), resulting in the following data set (concentrations in mg/m^3, and monitoring duration of 8 hr):

2.5 2.1 2.5 2.1 1.3 2.4 2.5 2.2 1.9
1.8 12.0 2.0 2.2 1.8 2.9 2.8 9.8

This data set is more easily interpreted if it is arranged in a rank order scheme as in Table II.1. The first column lists the relative rank. The second column gives the plotting position to

TABLE II.1
Rank Ordered Data for Analysis

r = Rank	r/(n+1) Plotting Position	Concentration mg/m^3	ln(Concentration)
1	0.056	1.3	0.262
2	0.111	1.8	0.588
3	0.167	1.8	0.588
4	0.222	1.9	0.642
5	0.278	2.0	0.693
6	0.333	2.1	0.742
7	0.389	2.1	0.742
8	0.444	2.2	0.788
9	0.500	2.2	0.788
10	0.556	2.4	0.875
11	0.611	2.5	0.916
12	0.667	2.5	0.916
13	0.722	2.5	0.916
14	0.778	2.8	1.030
15	0.833	2.9	1.065
16	0.889	9.8	2.282
17	0.944	12.0	2.485

plot the data on probability paper. For this example using 17 data points, the plotting position is given by the ratio: (rank/18.0) and is used in Figures II.1 and II.2. The third column lists the concentration data in mg/m^3. This is used for estimating descriptive statistics about the data (min, max, range, mode, median, mean, sample standard deviation). The fourth column lists the natural logarithm of the concentration data. It is used for calculating the geometric mean and the geometric standard deviation, the parameters that describe the data if it is lognormally distributed.

Figure II.3 shows the data from Column 3 of Table II.1 plotted as a histogram or probability density function. Each bin is 0.25 mg/m^3 wide.

Applying the concepts and equations of this section to the data displayed in Table II.1 and Figure II.3, descriptive statistics for the dust concentration data are quickly tabulated (Table II.2).

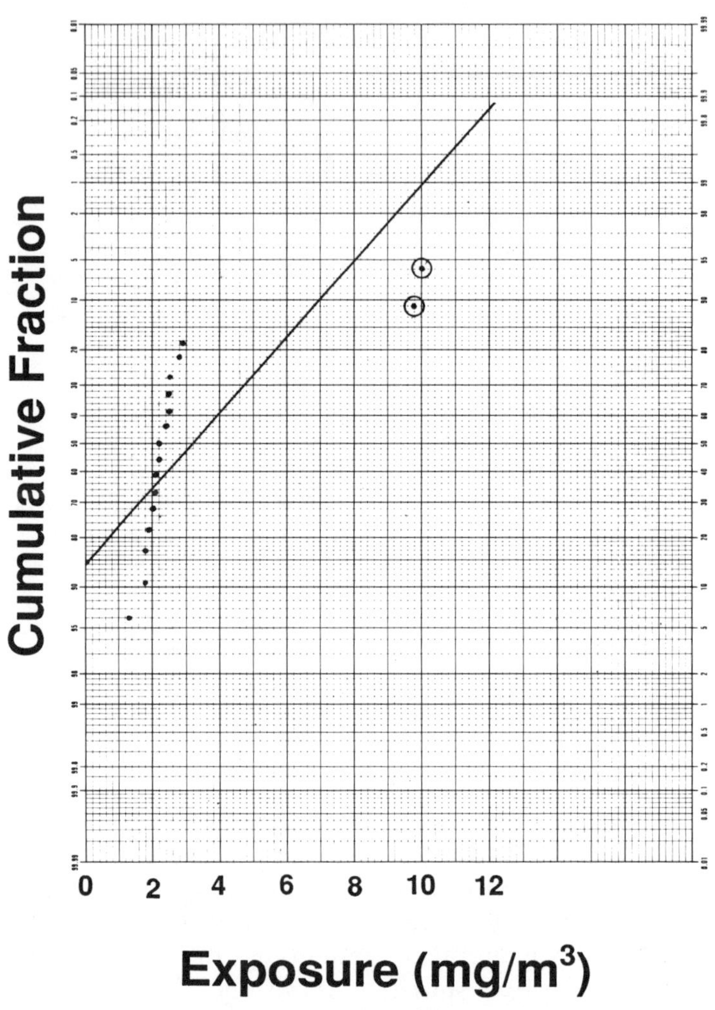

Figure II.1. Linear probability plot of dust exposure data

The minimum, maximum, and range are computed directly. The median is the value indicated as Rank 9 of Table II.1. The mode is estimated by inspection from Figure II.3, where the most frequent value is clearly between 2.0 and 2.25 mg/m^3. The mean and standard deviation are calculated from the data in Column 3, Table II.1. The geometric mean and geometric standard deviation are calculated from Column 4 of Table II.1. The intermediate results from use of Equations II.3 and II.4 are

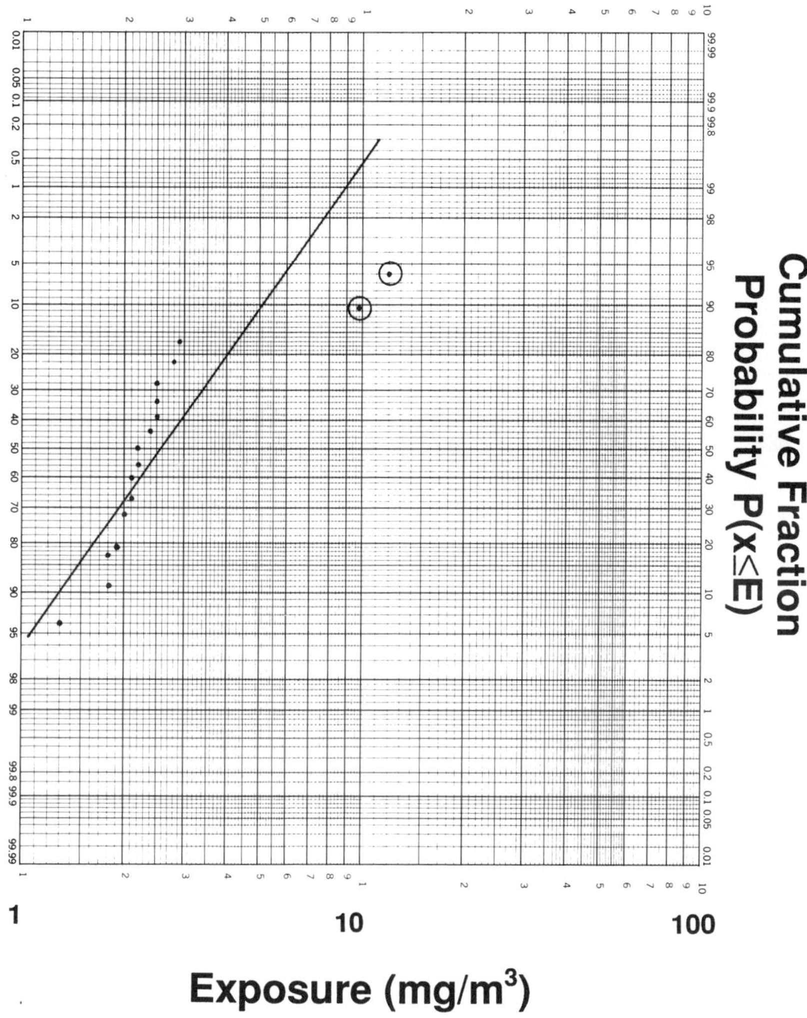

Figure II.2. Log probability plot of dust concentration data

tabulated to clarify calculation procedures here and in Appendixes II and III.

A note about units is in order. The standard deviation is the difference between the value of the cumulative probability distribution at the 84th percentile and its value at the 50th percentile on linear probability paper. As such, it has the same units as the data (mg/m^3 in this case). The geometric standard

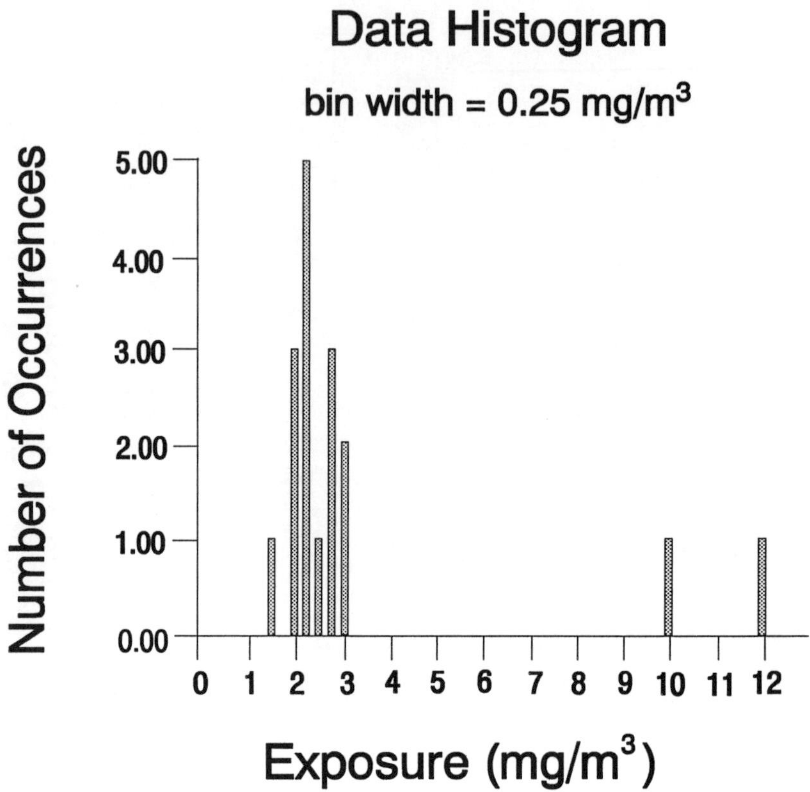

Figure II.3. Histogram of dust exposure data

deviation is the ratio of the value of the cumulative probability distribution at the 84th percentile and its value at the 50th percentile. As a ratio of two numbers having identical units, it has no units of its own.

Graphical Interpretation of Data

Probability distribution function plots are very powerful intuitive tools. They become even more powerful when the data are plotted on linear probability paper (normally distributed data) or log probability paper (lognormally distributed data). If all the data cluster closely to a straight line on one of these graphs, there is strong evidence that it all comes from a single

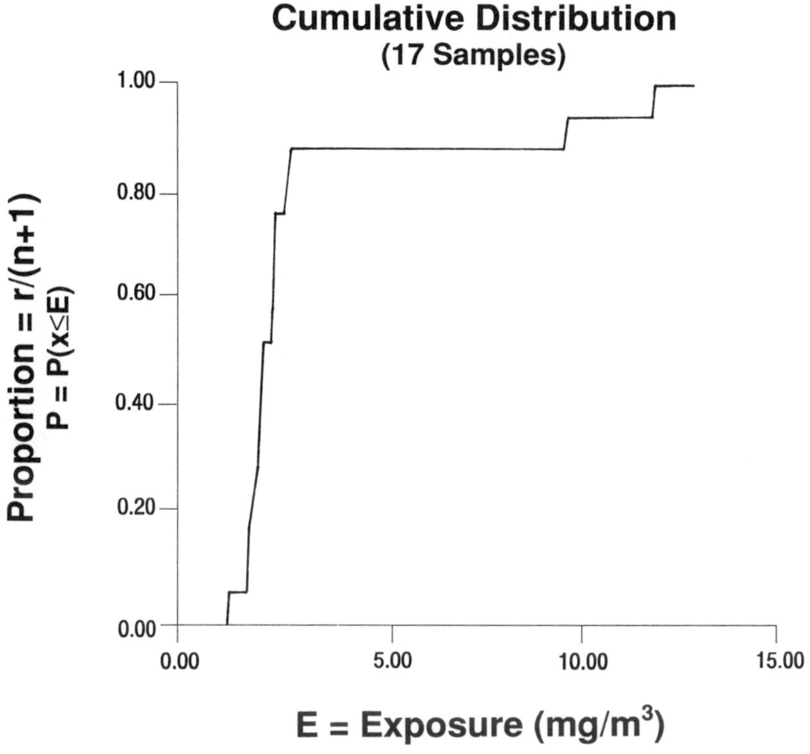

Figure II.4. Cumulative distribution plot of dust exposure data

population (the underlying homogeneous exposure group is well defined). If the data contain outliers that deserve special investigation, they stand out immediately.

In Figure II.4, the data from Table II.1 are plotted as a cumulative distribution function showing the proportion of exposures smaller than the plotted exposure. For example, 88% of the exposures are less than 9.8 mg/m^3.

From this plot it is apparent that 94% of the data are below the OEL (10 mg/m^3), 88% are below OEL/2, and none of the data are below OEL/10. Though this kind of plot provides useful insight, plotting exposure data on linear probability or log probability paper can be far more revealing. This allows a quick visual inspection to reveal the quality of fit to one or the other of the useful distributions.

TABLE II.2
Descriptive Statistics for Dust Exposures

Parameter Name	Symbol	Value	Units
Number of Samples	n	17	Unitless
Degrees of Freedom	υ	16	Unitless
Minimum	Min.	1.3	mg/m^3
Maximum	Max.	12.0	mg/m^3
Range	R	10.7	mg/m^3
Mode	Mo	2.13	mg/m^3
Median	Me	2.2	mg/m^3
Mean	M	3.2	mg/m^3
Standard Deviation	S	2.94	mg/m^3
Mean of [ln(concentration)]	ln(GM)	0.956	Unitless
Standard Deviation of [ln(concentration)]	ln(GS)	0.571	Unitless
Geometric Mean	GM	2.6	mg/m^3
Geometric Standard Deviation	GS	1.77	Unitless

Figures II.1 and II.2 show the data plotted on linear and log probability scales, respectively. This kind of plot is very convenient because if the data are parametrically distributed as either a normal or a lognormal distribution, they will plot as a straight line on one of these charts.

The straight line in Figure II.1 can be drawn to match the computed values of the mean, M, and the standard deviation, S, from Table II.2. One should note that it passes through the mean of the data at the 50th percentile and that the difference between the 84th and the 50th percentile is approximately equal to the standard deviation (as is the difference between the 50th and the 16th percentile). One should note also that the straight line does not fit the data well. Only if the two circled points were removed would a straight line be a reasonable fit to this data.

Figure II.2 shows the data plotted on log probability scales. This kind of plot is very convenient because if the data are lognormally distributed, it will plot as a straight line. The straight line is drawn to match the computed values of the geometric mean, GM, and the geometric standard deviation, GS, from Table II.2. One should note that the straight line

TABLE II.3
Rank Ordered Data for Analysis
(Outlying Data Points Removed)

r = Rank	r/(n+1) Plotting Position	Concentration mg/m³	ln(Concentration)
1	0.063	1.3	0.262
2	0.125	1.8	0.588
3	0.188	1.8	0.588
4	0.250	1.9	0.642
5	0.313	2.0	0.693
6	0.375	2.1	0.742
7	0.438	2.1	0.742
8	0.500	2.2	0.788
9	0.563	2.2	0.788
10	0.625	2.4	0.875
11	0.688	2.5	0.916
12	0.750	2.5	0.916
13	0.813	2.5	0.916
14	0.875	2.8	1.030
15	0.938	2.9	1.065

passes through the geometric mean of the data at the 50th percentile and that the geometric standard deviation is equal to the ratio of the value at the 84th percentile to the value at the 50th percentile (and is equal to the ratio of the values at the 50th to the 16th percentile).

In both Figures II.1 and II.2, the circled points seem not to fit the assumed distribution. In fact, the lines drawn seem not to fit any of the plotted points.* This is a strong indication that the data in hand do not represent a single homogeneous exposure group. Since the 15 smallest exposures seem to plot as a straight line, one might reasonably assume that they do represent

This is a qualitative indication of poor fit to the assumed distribution. Quantitative goodness-of-fit testing is discussed in the Dupont copyrighted LOGAN software and manual available from the American Industrial Hygiene Association (AIHA).

homogeneous exposures. If investigation reveals the two highest exposures to be members of a different population (process upset, procedures violated, etc.), the two highest exposures should be removed from the data set. This is an example of using descriptive statistics to improve professional judgment.

Missing Data

One of the great benefits of probability plotting is that the distribution of data can be modeled even when some of the data are unusable. If, for example, the detection limit for total dust is 1.8 mg/m^3, this would make the first three samples listed in Table II.1 and plotted in Figures II.1 and II.2 unplottable. In addition, if an investigation of samples ranked 16 and 17 revealed an analytical quality assurance problem, they would be rendered unplottable. By leaving these five samples unplotted and by plotting the remaining 12 samples in their proper rank order position, sketching a straight line defining the distribution would still be possible.

The value of the straight line at the 16th, 50th, and 84th percentile is then used to compute the distributional parameters.

On linear probability paper, the mean is the point where the plotted line crosses the 50th percentile, and the standard deviation is half the difference between the values at the 84th and the 16th percentiles. This analysis applies if the data fit a normal distribution.

On log probability paper, the geometric mean is the point where the plotted line crosses the 50th percentile, and the geometric standard deviation is the square root of the ratio of the values at the 84th and the 16th percentiles. This analysis applies if the data fit a lognormal distribution.

Defining a Smaller HEG

Suppose an investigation revealed that the analysis was proper and that the local exhaust system had malfunctioned during the collection of data ranked 16 and 17 in Table II.1. Clearly, these two data points would represent different conditions than the other 15 points. They should be analyzed to define conditions expected when the exhaust system is off. Analyzing the 15 valid samples together to define conditions expected for the homogeneous exposure group under the normal working conditions with the exhaust system functioning is important. This requires a recalculation of the plotting positions as shown in Table II.3 and of descriptive statistics as shown in Table II.4.

Figures II.5 and II.6 show the linear probability and log probability plots of the data, respectively. These data fit a straight line reasonably well on both kinds of paper. There is no obvious reason to choose either the normal or the lognormal as the preferred distribution. In this case, a closer evaluation of the data is required to make a distributional choice.

Evaluating the Parametric Assumption

Some very simple observations can be applied to descriptive statistics to evaluate the quality of the assumption that a data set is well represented by either normal or lognormal parameters.

TABLE II.4
Descriptive Statistics for Dust Exposures
15 Data Points, Outliers Removed

Parameter Name	Symbol	Value	Units
Number of Samples	n	15	Unitless
Degrees of Freedom	υ	14	Unitless
Minimum	Min.	1.3	mg/m^3
Maximum	Max.	2.9	mg/m^3
Range	R	1.6	mg/m^3
Mode	Mo	2.13	mg/m^3
Median	Me	2.2	mg/m^3
Mean	M	2.20	mg/m^3
Standard Deviation	S	0.417	mg/m^3
Mean of [ln(concentration)]	ln(GM)	0.770	Unitless
Standard Deviation of [ln(concentration)]	ln(GS)	0.203	Unitless
Geometric Mean	GM	2.16	mg/m^3
Geometric Standard Deviation	GS	1.225	Unitless

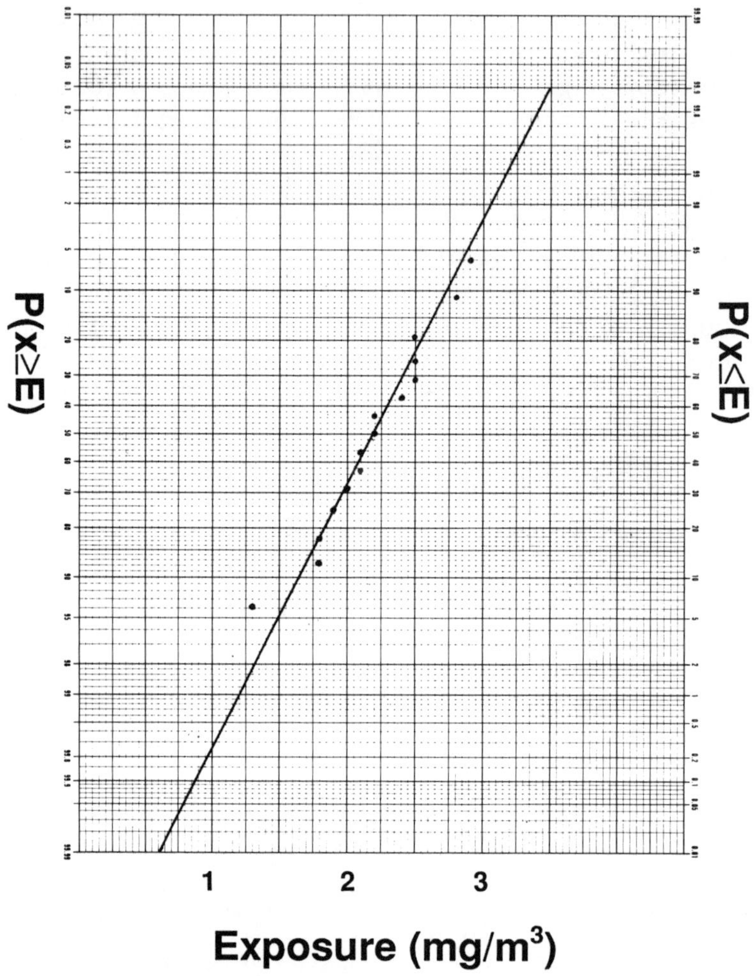

Figure II.5. Normal probability plot of 15 dust exposure samples

A normal distribution is symmetrical and the mode, median, and mean are equal and have a value of μ. When the data are symmetrical (straight line on linear probability plot) and Mo = Me = M, the normal assumption is quite robust.

The two-parameter lognormal distribution is asymmetrical and has no negative values. Its mode is less than its median, which is less than its mean. The spacing between these three parameters is related to the size of the geometric standard deviation. Large GS implies a large amount of skew and wide

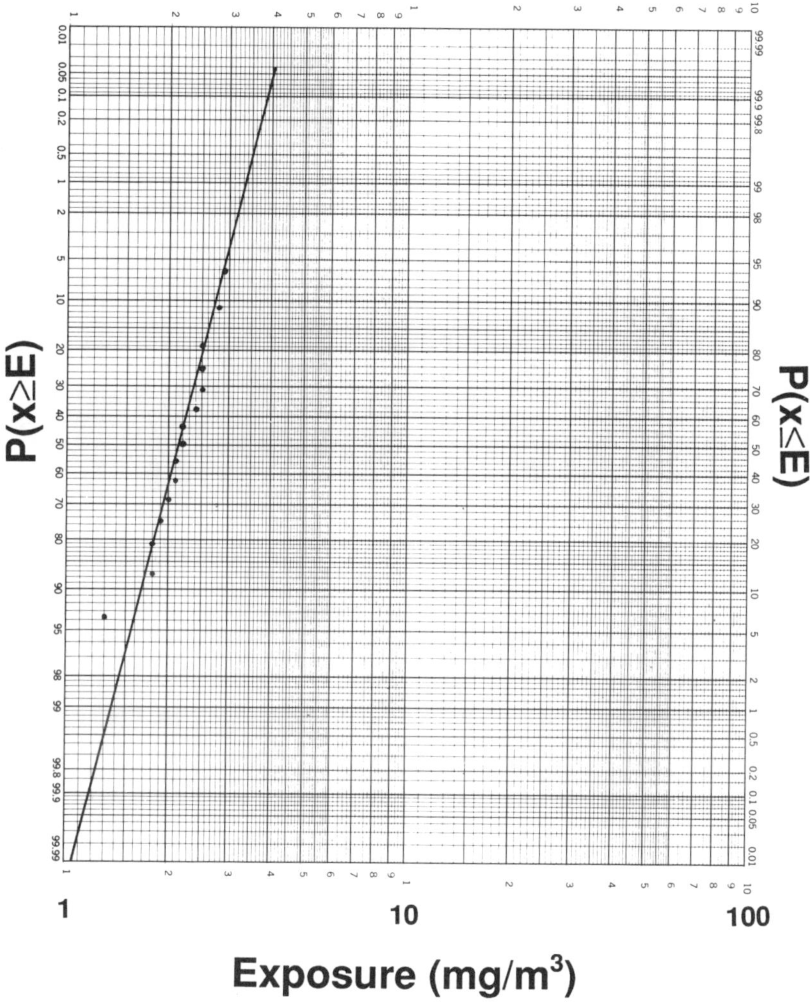

Figure II.6. Lognormal probability plot of 15 exposure samples

separation with the Mo < Me < M, as in Equations II.5, II.6, and II.7.

$$\text{Mo} = \exp\{\ln(\text{GM}) - [\ln(\text{GS})]^2\} \tag{II.5}$$

$$\begin{aligned}\text{Me} &= \exp[\ln(\text{GM})] \\ &= \text{GM}\end{aligned} \tag{II.6}$$

TABLE II.5
Cumulative Normal Distribution — Values of P^A

z_p	.00	.01	.02	.03	.04	.05	.06	.07	.08	.09
.0	.5000	.5040	.5080	.5120	.5160	.5199	.5239	.5279	.5319	.5359
.1	.5398	.5438	.5478	.5517	.5557	.5596	.5636	.5675	.5714	.5753
.2	.5793	.5832	.5871	.5910	.5948	.5987	.6026	.6064	.6103	.6141
.3	.6179	.6217	.6255	.6293	.6331	.6368	.6406	.6443	.6480	.6517
.4	.6554	.6591	.6628	.6664	.6700	.6736	.6772	.6808	.6844	.6879
.5	.6915	.6950	.6985	.7019	.7054	.7088	.7123	.7157	.7190	.7224
.6	.7257	.7291	.7324	.7357	.7389	.7422	.7454	.7486	.7517	.7549
.7	.7580	.7611	.7642	.7673	.7704	.7734	.7764	.7794	.7823	.7852
.8	.7881	.7910	.7939	.7967	.7995	.8023	.8051	.8078	.8106	.8133
.9	.8159	.8186	.8212	.8238	.8264	.8289	.8315	.8340	.8365	.8389
1.0	.8413	.8438	.8461	.8485	.8508	.8531	.8554	.8577	.8599	.8621
1.1	.8643	.8665	.8686	.8708	.8729	.8749	.8770	.8790	.8810	.8830
1.2	.8849	.8869	.8888	.8907	.8925	.8944	.8962	.8980	.8997	.9015
1.3	.9032	.9049	.9066	.9082	.9099	.9115	.9131	.9147	.9162	.9177
1.4	.9192	.9207	.9222	.9236	.9251	.9265	.9279	.9292	.9306	.9319
1.5	.9332	.9345	.9357	.9370	.9382	.9394	.9406	.9418	.9429	.9441
1.6	.9452	.9463	.9474	.9484	.9495	.9505	.9515	.9525	.9535	.9545
1.7	.9554	.9564	.9573	.9582	.9591	.9599	.9608	.9616	.9625	.9633
1.8	.9641	.9649	.9656	.9664	.9671	.9678	.9686	.9693	.9699	.9706
1.9	.9713	.9719	.9726	.9732	.9738	.9744	.9750	.9756	.9761	.9767
2.0	.9772	.9778	.9783	.9788	.9793	.9798	.9803	.9808	.9812	.9817
2.1	.9821	.9826	.9830	.9834	.9838	.9842	.9846	.9850	.9854	.9857
2.2	.9861	.9864	.9868	.9871	.9875	.9878	.9881	.9884	.9887	.9890
2.3	.9893	.9896	.9898	.9901	.9904	.9906	.9909	.9911	.9913	.9916
2.4	.9918	.9920	.9922	.9925	.9927	.9929	.9931	.9932	.9934	.9936
2.5	.9938	.9940	.9941	.9943	.9945	.9946	.9948	.9949	.9951	.9952
2.6	.9953	.9955	.9956	.9957	.9959	.9960	.9961	.9962	.9963	.9964
2.7	.9965	.9966	.9967	.9968	.9969	.9970	.9971	.9972	.9973	.9974
2.8	.9974	.9975	.9976	.9977	.9977	.9978	.9979	.9979	.9980	.9981
2.9	.9981	.9982	.9982	.9983	.9984	.9984	.9985	.9985	.9986	.9986
3.0	.9987	.9987	.9987	.9988	.9988	.9989	.9989	.9989	.9990	.9990
3.1	.9990	.9991	.9991	.9991	.9992	.9992	.9992	.9992	.9993	.9993
3.2	.9993	.9993	.9994	.9994	.9994	.9994	.9994	.9995	.9995	.9995
3.3	.9995	.9995	.9995	.9996	.9996	.9996	.9996	.9996	.9996	.9997
3.4	.9997	.9997	.9997	.9997	.9997	.9997	.9997	.9997	.9997	.9998

[A]Values of P corresponding to z_p for the normal curve. z is the standard normal variable. The value of P for $-z_p$ equals one minus the value of P for $+z_p$, e.g., the P for −1.62 equals 1 − .9474 = .0526.

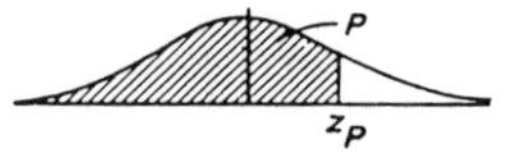

TABLE II.6
Cumulative Normal Distribution — Values of z_p [A]

P	.00	.01	.02	.03	.04	.05	.06	.07	.08	.09
.00	—	−2.33	−2.05	−1.88	−1.75	−1.64	−1.55	−1.48	−1.41	−1.34
.10	−1.28	−1.23	−1.18	−1.13	−1.08	−1.04	−0.99	−0.95	−0.92	−0.88
.20	−0.84	−0.81	−0.77	−0.74	−0.71	−0.67	−0.64	−0.61	−0.58	−0.55
.30	−0.52	−0.50	−0.47	−0.44	−0.41	−0.39	−0.36	−0.33	−0.31	−0.28
.40	−0.25	−0.25	−0.20	−0.18	−0.15	−0.13	−0.10	−0.08	−0.05	−0.03
.50	0.00	0.03	0.05	0.08	0.10	0.13	0.15	0.18	0.20	0.23
.60	0.25	0.28	0.31	0.33	0.36	0.39	0.41	0.44	0.47	0.50
.70	0.52	0.55	0.58	0.61	0.64	0.67	0.71	0.74	0.77	0.81
.80	0.84	0.88	0.92	0.95	0.99	1.04	1.08	1.13	1.18	1.23
.90	1.28	1.34	1.41	1.48	1.55	1.64	1.75	1.88	2.05	2.33

Special Values

P	.001	.005	.010	.025	.050	.100
z_p	−3.090	−2.576	−2.326	−1.960	−1.645	−1.282

P	.999	.995	.990	.975	.950	.900
z_p	3.090	2.576	2.326	1.960	1.645	1.282

[A]Values of z_p corresponding to P for the normal curve. z is the standard normal variable.

$$M = \exp\{\ln(GM) + (1/2)[\ln(GS)]^2\} \tag{II.7}$$

One should note that the median is equal to the geometric mean. This is often a good discriminator of the quality of the distributional assumption. For example, the median is unequal to the geometric mean in Table II.2, and the data do not plot as a straight line in Figure II.2. This suggests the lognormal distribution is a poor choice to represent this data. In contrast, the median equals the geometric mean in Table II.4, and the data plot as a straight line in Figure II.6. This suggests that without the outliers, the remaining 15 data points may be well represented by a lognormal distribution.

Equations II.8 and II.9 provide a useful relationship between parameters of a fitted distribution and values of the data in the distribution. This technique is useful in many calculations involving parametrically distributed data.† Let z_p be the standard normal deviate for the pth percentile of the standard normal distribution.

The value of the data, x_p, at the pth percentile of a normal distribution is given by:

$$x_p = M + (z_p)(S) \qquad \textbf{(II.8)}$$

The value of the data, x_p, at the pth percentile of a lognormal distribution is given by:

$$x_p = \exp\{\ln(GM) + (z_p)[\ln(GS)]\} \qquad \textbf{(II.9)}$$

The relation between p and z_p is given by Tables II.5 and II.6. These tables plus Equations II.8 and II.9 form the basis for the probability plots described earlier in this appendix.

SUMMARY

This section briefly reviews the theme of Appendix II and illustrates its conclusions one last time. The theme is that statistical analysis of a group of data is best conducted by using a mixture of graphical and analytical tools. The descriptive statistics are easily calculated for every data set, whether it fits a distribution or not. Likewise, every distribution is easily plotted on probability paper. If the data do not fit a straight line, they are not suited for analysis with parametric statistics—even though the parameters have already been calculated.

The descriptive statistics (for the 17-measurement HEG) contained in Table II.2 show clearly that the mean, median, and mode are not equal. This suggests that the data are not well represented by a normal distribution. The mode, median, and mean are correctly ranked for a lognormal distribution, suggesting that it may be the better model for this data. However, the median is unequal to the geometric mean, suggesting a deviation from true lognormality. This quantitative evaluation thus agrees with the conclusion based on the nonlinearity of data plotted in Figures II.1 and II.2. Figures II.7

---

†In one specific application, one should note that $z_p = 1$ when $p = 0.84$, and $z_p = 0$ when $p = 0.50$. The reader is encouraged to prove that ($GS = x_{0.84}/x_{0.5}$), as was described in the section on Missing Data.

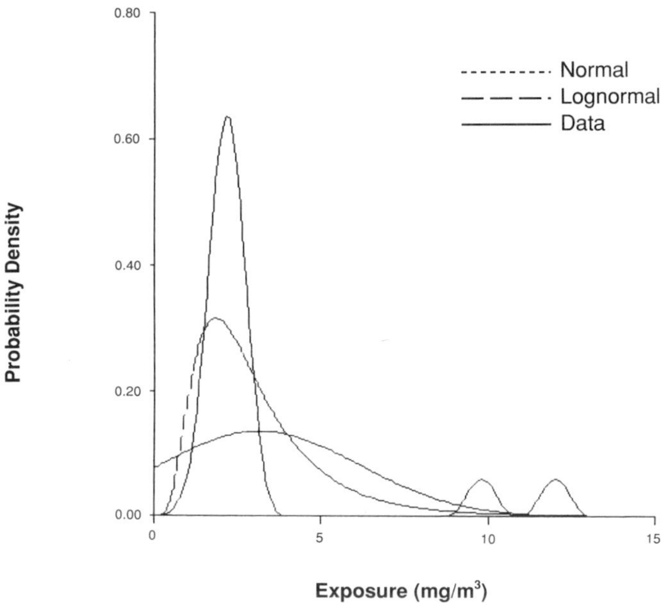

Figure II.7. Probability density functions for the 17-member exposure data set

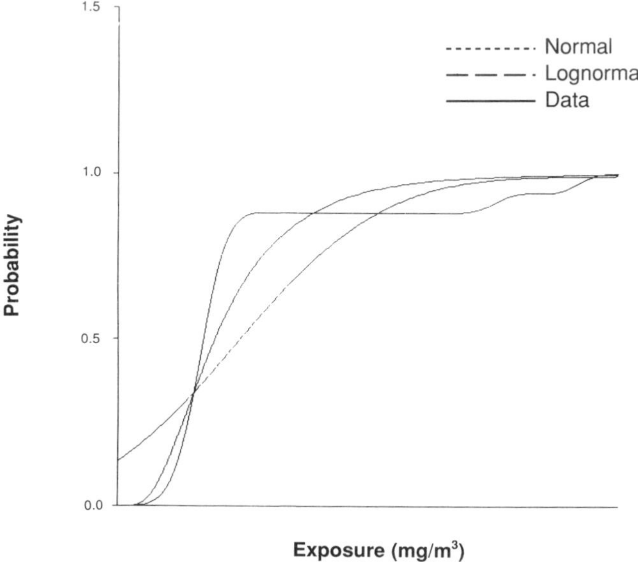

Figure II.8. Cumulative probability distribution function for the 17-member exposure data set

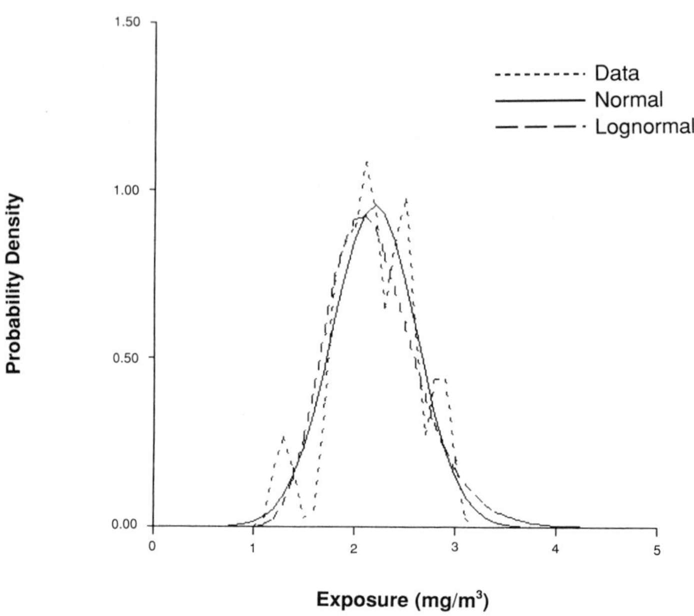

Figure II.9. Probability density function for 15-member exposure data set

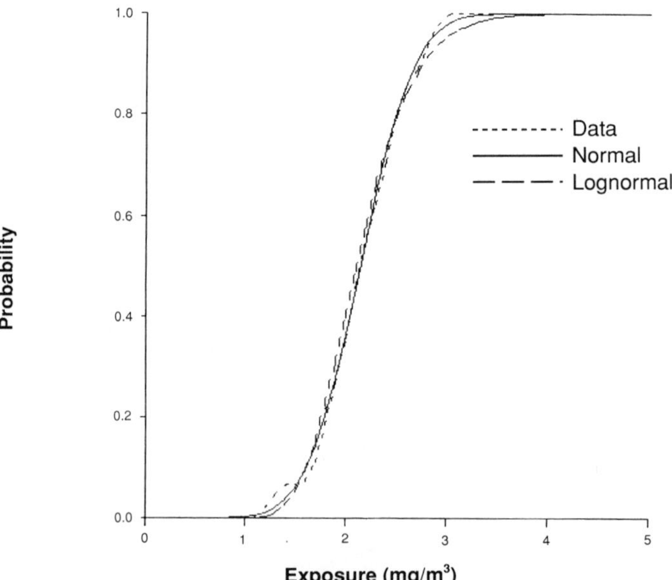

Figure II.10. Cumulative probability distribution function for 15-member exposure data set

and II.8 further illustrate the poor fit of these data to either the normal or the lognormal distribution.

In contrast, when one considers the descriptive statistics of the reduced data set (the 15-member HEG) in Table II.4, one notices that the geometric mean matches the median value found in Row 8 of Table II.3. This suggests lognormality. The mean, median, and mode are approximately equal, suggesting normality. Figures II.9 and II.10 confirm the goodness of fit for both the normal and the lognormal distributions. The choice between normality and lognormality is difficult, but because there are no negative data (and no possibility of negative data), the lognormal distribution is the better fit.

After working through a few of these problems an industrial hygienist can develop the needed intuitive sense of when data are adequately modeled by a normal or lognormal distribution assumption, that is, a sense of when to trust the parametric tests illustrated in Appendixes III and IV.

Appendix III
Arithmetic Mean Tests

OVERVIEW

The true population mean (arithmetic average of all exposures) represents the expected exposure of all workers in a homogeneous exposure group, *no matter the underlying distribution of the data*. The population mean can be estimated from sampled data. Appendix III illustrates techniques for estimating the population mean when the sampled data are normally or lognormally distributed. When the sampled data fit a normal distribution, the test is applied directly to the measured exposure data. When the sampled data are skewed so that they fit a lognormal distribution, the test is applied to the logarithm of the measured exposure data and results in logarithms of the upper and lower confidence limits and the logarithm of the mean and median.

The intent of Appendixes III and IV is to illustrate useful procedures rather than to provide a tutorial introduction to these procedures. Every reader is warned to avoid applying these tests blindly. Readers should learn the assumptions underlying these tests well or obtain the services of a professional statistician to ensure validity of the approach for each data set. Underlying assumptions, including stationary distribution, random data collection, unbiased monitoring and analytical procedures, and goodness of fit to assumed parametric distributions, must be verified.

Prior to applying inferential tests, the procedures in Appendix II should have been considered and applied where appropriate. Appendix II should be read before Appendixes III and IV. Appendixes II, III, and IV should be read in that order, as a unit, before the tools in any one appendix are used.

The 95% confidence interval on the mean is that range of values that will include the sample mean 95 times out of 100 when the specified number of samples is drawn from a population. In Appendix II, there are 17 samples available from which to evaluate the population of all exposures. The confidence interval on the mean enhances professional judgment by clarifying the precision of the estimate of the population mean.

APPROACH 1: USING NORMAL STATISTICS (CENTRAL LIMIT THEOREM)

When the data are symmetrically distributed and not extremely skewed (as in a lognormal distribution with a large geometric standard deviation [GS]), this approach is very robust. It becomes even more so if there is a large number of samples (n > 30). This test uses the Student *t* distribution and has as its assumption that the sample data are all that is known about the population. If there is independent knowledge of the variance of the population, stronger inferences can be drawn by using the standard normal variate, *z*, in place of the Student variate, *t*.*

The goal of this calculation is a two-sided confidence interval on the mean with 95% confidence. The upper confidence limit should be the upper bound of 97.5% of the area under the Student *t* distribution, and the lower confidence limit should be the lower bound below 97.5% of its area. There should be 2.5% of the area in each tail.

Descriptive statistics from Table II.2 should be used: M = 3.2 mg/m^3, S = 2.9 mg/m^3, n = 17 exposure measurements, and $\upsilon = 16$ degrees of freedom. If one examines a Student's *t* distribution table (as shown in Table III.1), one will find that $t_{0.975,16} = 2.120$ (the one-sided *t* value for determining a 95% confidence interval with 16 degrees of freedom). The upper and lower confidence limits on the mean should be computed with Equations III.1 and III.2.

*For large sample sizes, n > 120, the *t* variate is approximately equal to the *z* variate (as shown in Table III.1).

TABLE III.1
Percentiles of the *t* Distribution[A]

df	$t_{.60}$	$t_{.70}$	$t_{.80}$	$t_{.90}$	$t_{.95}$	$t_{.975}$	$t_{.99}$	$t_{.995}$
1	.325	.727	1.376	3.078	6.314	12.706	31.821	63.657
2	.289	.617	1.061	1.886	2.920	4.303	6.965	9.925
3	.277	.584	.978	1.638	2.353	3.182	4.541	5.841
4	.271	.569	.941	1.533	2.132	2.776	3.747	4.604
5	.267	.559	.920	1.476	2.015	2.571	3.365	4.032
6	.265	.553	.906	1.440	1.943	2.447	3.143	3.707
7	.263	.549	.896	1.415	1.895	2.365	2.998	3.499
8	.262	.546	.889	1.397	1.860	2.306	2.896	3.355
9	.261	.543	.883	1.383	1.833	2.262	2.821	3.250
10	.260	.542	.879	1.372	1.812	2.228	2.764	3.169
11	.260	.540	.876	1.363	1.796	2.201	2.718	3.106
12	.259	.539	.873	1.356	1.782	2.179	2.681	3.055
13	.259	.538	.870	1.350	1.771	2.160	2.650	3.012
14	.258	.537	.868	1.345	1.761	2.145	2.624	2.977
15	.258	.536	.866	1.341	1.753	2.131	2.602	2.947
16	.258	.535	.865	1.337	1.746	2.120	2.583	2.921
17	.257	.534	.863	1.333	1.740	2.110	2.567	2.898
18	.257	.534	.862	1.330	1.734	2.101	2.552	2.878
19	.257	.533	.861	1.328	1.729	2.093	2.539	2.861
20	.257	.533	.860	1.325	1.725	2.086	2.528	2.845
21	.257	.532	.859	1.323	1.721	2.080	2.518	2.831
22	.256	.532	.858	1.321	1.717	2.074	2.508	2.819
23	.256	.532	.858	1.319	1.714	2.069	2.500	2.807
24	.256	.531	.857	1.318	1.711	2.064	2.492	2.797
25	.256	.531	.856	1.316	1.708	2.060	2.485	2.787
26	.256	.531	.856	1.315	1.706	2.056	2.479	2.779
27	.256	.531	.855	1.314	1.703	2.052	2.473	2.771
28	.256	.530	.855	1.313	1.701	2.048	2.467	2.763
29	.256	.530	.854	1.311	1.699	2.045	2.462	2.756
30	.256	.530	.854	1.310	1.697	2.042	2.457	2.750
40	.255	.529	.851	1.303	1.684	2.021	2.423	2.704
60	.254	.527	.848	1.296	1.671	2.000	2.390	2.660
120	.254	.526	.845	1.289	1.658	1.980	2.358	2.617
∞	.253	.524	.842	1.282	1.645	1.960	2.326	2.576

[A]From National Bureau of Standards (refer to bibliography for appendixes).

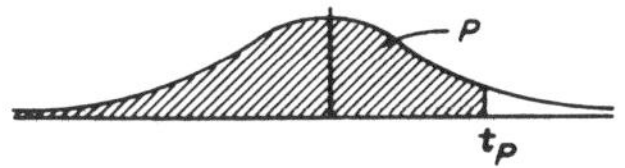

$$\begin{aligned}\text{UCL} &= \text{M} + \left(t_{0.975,16}\right)\left(\frac{\text{S}}{\sqrt{\text{n}}}\right)\\ &= 3.2 + (2.12)\left(\frac{2.9}{\sqrt{17}}\right)\\ &= 4.7\ \text{mg/m}^3\end{aligned} \tag{III.1}$$

$$\begin{aligned}\text{LCL} &= \text{M} - \left(t_{0.975,16}\right)\left(\frac{\text{S}}{\sqrt{\text{n}}}\right)\\ &= 3.2 - (2.12)\left(\frac{2.9}{\sqrt{17}}\right)\\ &= 1.7\ \text{mg/m}^3\end{aligned} \tag{III.2}$$

Approach 2: Using Lognormal Statistics

When the data are skewed and fit a lognormal distribution, the logarithmic transformation may be applied to create a data set that is normally distributed. The normally distributed, log-transformed data may be analyzed using the procedures described in the section on Approach 1: Using Normal Statistics (Central Limit Theorem) in this appendix. The calculation results in log-transformed upper and lower confidence limits on the median.

In exposure monitoring applications, the mean is the variable of interest, not the median. From Equation II.7 the mean is a function of both ln(GM) and of ln(GS). Under the reasonable assumption that the distribution of estimates of ln(GM) will match the distribution of estimates of ln(M), the confidence interval on the mean is computed by adding and subtracting the t factor to ln(M).

The number of samples, n=17, the number of statistical degrees of freedom, $\upsilon = 16$, and the value of $(t_{0.975,16}) = 2.120$ all remain the same as in the section of this appendix describing Approach 1. The descriptive statistics from Table II.2 can be used with Equation II.7 (shown here) to calculate the numerical value for ln(M).

$$\begin{aligned} \ln(M) &= \ln(GM) + (0.5)(\ln(GS))^2 \\ &= 0.956 + (0.5)(0.571)^2 \\ &= 1.119 \\ M &= 3.06 \text{ mg/m}^3 \end{aligned} \quad \textbf{(II.7)}$$

Then, the log-transformed analogues of equations III.1 and III.2 should be used to estimate the confidence interval.

$$\begin{aligned} \ln(GUCL) &= \ln(M) + \left(t_{0.975,16}\right)\left(\frac{\ln(GS)}{\sqrt{n}}\right) \\ &= 1.119 + (2.12)\left(\frac{0.571}{\sqrt{17}}\right) \\ &= 1.41 \\ GUCL &= 4.10 \text{ mg/m}^3 \end{aligned} \quad \textbf{(III.3)}$$

$$\begin{aligned} \ln(GLCL) &= \ln(M) - \left(t_{0.975,16}\right)\left(\frac{\ln(GS)}{\sqrt{n}}\right) \\ &= 1.119 - (2.12)\left(\frac{0.571}{\sqrt{17}}\right) \\ &= 0.825 \\ GLCL &= 2.28 \text{ mg/m}^3 \end{aligned} \quad \textbf{(III.4)}$$

SUMMARY OF APPROACHES 1 AND 2 FOR THE EXAMPLE

The numerical illustrations were based on descriptive statistics from Table II.2 even though those data are not well represented by the parametric assumption. The calculations in Equations III.1 through III.4 can be easily rerun for the refined homogeneous exposure group represented by descriptive statistics from Table II.4.

TABLE III.2
Confidence Interval Summary

Assumed Distribution	n^A	OEL^B mg/m^3	LCL^C mg/m^3	M^D mg/m^3	UCL^E mg/m^3
Normal	17	10	1.70	3.20	4.70
Lognormal	17	10	2.28	3.06	4.10
Normal	15	10	1.97	2.200	2.43
Lognormal	15	10	1.93	2.205	3.42

[A]n = Number of samples.
[B]OEL = Occupational exposure limit.
[C]LCL = Lower confidence limit.
[D]M = Mean.
[E]UCL = Upper confidence limit.

The reader is encouraged to do these calculations and to compare results with the summary found in Table III.2. From Table II.4, the reader should use n = 15, $\upsilon = 14$, ln(GM) = 0.770, ln(GS) = 0.203, and should find $t_{0.975,14} = 2.145$ from Table III.1.

In both data sets, the estimate of the mean computed by averaging compares very closely with the estimate computed by a fitted lognormal distribution. This is especially noteworthy in the case of the 17-member sample since the data do not fit the presumed distribution well in that case.

In the 15-sample data set, there is negligible difference between the confidence intervals estimated under the normal and lognormal assumption. This is to be expected since the data fit both distributions well.

What conclusion can be drawn from this analysis? It demonstrates 95% confidence that the average value of all exposures in the homogeneous exposure group is less than half the occupational exposure limit since none of the estimates of the upper confidence limit exceed 5.0 mg/m^3.

Appendix IV
Tolerance Limit Test

Overview

There are acute hazards that do not correlate well with the mean of exposures in a homogeneous exposure group. For these hazards, the workplace evaluation is based on extreme values, not average exposures. A commonly used measure of merit is to require 90% of exposures to be below half the occupational exposure limit (OEL). One approach to this question is simply to use the 90th percentile exposure from the probability plot. This works fine as a descriptor of the sample data.

However, the mean and standard deviation evident in the probability plot are estimated from sample data and seldom equal the population mean and standard deviation. The uncertainty in these estimates results in uncertainty in the vertical position of the probability plot (in the amount of the confidence interval around the mean) and uncertainty in the slope of the probability plot (in the amount of the confidence interval around the variance). Together, the uncertainty in position and slope both contribute to the uncertainty in the point where the true population distribution crosses the percentile of interest. The tolerance interval is a calculation that reveals the amount of uncertainty in an estimate of a percentile of the distribution of exposures.

The intent of Appendixes III and IV is to illustrate useful procedures rather than provide a tutorial introduction to these procedures. Every reader is warned to avoid applying these tests blindly. Readers should learn the assumptions underlying these tests well or obtain the services of a professional statistician to ensure validity of the approach for each data set. Underlying assumptions, including stationary distribution,

TABLE IV.1
Factors for One-Sided Tolerance Limits for Normal Distributions[A,B]

	γ = 0.75					γ = 0.90				
n \ P	0.75	0.90	0.95	0.99	0.999	0.75	0.90	0.95	0.99	0.999
3	1.464	2.501	3.152	4.396	5.805	2.602	4.258	5.310	7.340	9.651
4	1.256	2.134	2.680	3.726	4.910	1.972	3.187	3.957	5.437	7.128
5	1.152	1.961	2.463	3.421	4.507	1.698	2.742	3.400	4.666	6.112
6	1.087	1.860	2.336	3.243	4.273	1.540	2.494	3.091	4.242	5.556
7	1.043	1.791	2.250	3.126	4.118	1.435	2.333	2.894	3.972	5.201
8	1.010	1.740	2.190	3.042	4.008	1.360	2.219	2.755	3.783	4.955
9	0.984	1.702	2.141	2.977	3.924	1.302	2.133	2.649	3.641	4.772
10	0.964	1.671	2.103	2.927	3.858	1.257	2.065	2.568	3.532	4.629
11	0.947	1.646	2.073	2.885	3.804	1.219	2.012	2.503	3.444	4.515
12	0.933	1.624	2.048	2.851	3.760	1.188	1.966	2.448	3.371	4.420
13	0.919	1.606	2.026	2.822	3.722	1.162	1.928	2.403	3.310	4.341
14	0.909	1.591	2.007	2.796	3.690	1.139	1.895	2.363	3.257	4.274
15	0.899	1.577	1.991	2.776	3.661	1.119	1.866	2.329	3.212	4.215
16	0.891	1.566	1.977	2.756	3.637	1.101	1.842	2.299	3.172	4.164
17	0.883	1.554	1.964	2.739	3.615	1.085	1.820	2.272	3.136	4.118
18	0.876	1.544	1.951	2.723	3.595	1.071	1.800	2.249	3.106	4.078
19	0.870	1.536	1.942	2.710	3.577	1.058	1.781	2.228	3.078	4.041
20	0.865	1.528	1.933	2.697	3.561	1.046	1.765	2.208	3.052	4.009
21	0.859	1.520	1.923	2.686	3.545	1.035	1.750	2.190	3.028	3.979
22	0.854	1.514	1.916	2.675	3.532	1.025	1.736	2.174	3.007	3.952
23	0.849	1.508	1.907	2.665	3.520	1.016	1.724	2.159	2.987	3.927
24	0.845	1.502	1.901	2.656	3.509	1.007	1.712	2.145	2.969	3.904
25	0.842	1.496	1.895	2.647	3.497	0.999	1.702	2.132	2.952	3.882
30	0.825	1.475	1.869	2.613	3.454	0.966	1.657	2.080	2.884	3.794
35	0.812	1.458	1.849	2.588	3.421	0.942	1.623	2.041	2.833	3.730
40	0.803	1.445	1.834	2.568	3.395	0.923	1.598	2.010	2.793	3.679
45	0.795	1.435	1.821	2.552	3.375	0.908	1.577	1.986	2.762	3.638
50	0.788	1.426	1.811	2.538	3.358	0.894	1.560	1.965	2.735	3.604

TABLE IV.1 (Continued)
Factors for One-Sided Tolerance Limits for Normal Distributions[A,B]

	$\gamma = 0.95$					$\gamma = 0.99$				
n \ P	0.75	0.90	0.95	0.99	0.999	0.75	0.90	0.95	0.99	0.999
3	3.804	6.158	7.655	10.552	13.857	—	—	—	—	—
4	2.619	4.163	5.145	7.042	9.215	—	—	—	—	—
5	2.149	3.407	4.202	5.741	7.501	—	—	—	—	—
6	1.895	3.006	3.707	5.062	6.612	2.849	4.408	5.409	7.334	9.550*
7	1.732	2.755	3.399	4.641	6.061	2.490	3.856	4.730	6.411	8.348
8	1.617	2.582	3.188	4.353	5.686	2.252	3.496	4.287	5.811	7.566
9	1.532	2.454	3.031	4.143	5.414	2.085	3.242	3.971	5.389	7.014
10	1.465	2.355	2.911	3.981	5.203	1.954	3.048	3.739	5.075	6.603
11	1.411	2.275	2.815	3.852	5.036	1.854	2.897	3.557	4.828	6.284
12	1.366	2.210	2.736	3.747	4.900	1.771	2.773	3.410	4.633	6.032
13	1.329	2.155	2.670	3.659	4.787	1.702	2.677	3.290	4.472	5.826
14	1.296	2.108	2.614	3.585	4.690	1.645	2.592	3.189	4.336	5.651
15	1.268	2.068	2.566	3.520	4.607	1.596	2.521	3.102	4.224	5.507
16	1.242	2.032	2.523	3.463	4.534	1.553	2.458	3.028	4.124	5.374
17	1.220	2.001	2.486	3.415	4.471	1.514	2.405	2.962	4.038	5.268
18	1.200	1.974	2.453	3.370	4.415	1.481	2.357	2.906	3.961	5.167
19	1.183	1.949	2.423	3.331	4.364	1.450	2.315	2.855	3.893	5.078
20	1.167	1.926	2.396	3.295	4.319	1.424	2.275	2.807	3.832	5.003
21	1.152	1.905	2.371	3.262	4.276	1.397	2.241	2.768	3.776	4.932
22	1.138	1.887	2.350	3.233	4.238	1.376	2.208	2.729	3.727	4.866
23	1.126	1.869	2.329	3.206	4.204	1.355	2.179	2.693	3.680	4.806
24	1.114	1.853	2.309	3.181	4.171	1.336	2.154	2.663	3.638	4.755
25	1.103	1.838	2.292	3.158	4.143	1.319	2.129	2.632	3.601	4.706
30	1.059	1.778	2.220	3.064	4.022	1.249	2.029	2.516	3.446	4.508
35	1.025	1.732	2.166	2.994	3.934	1.195	1.957	2.431	3.334	4.364
40	0.999	1.697	2.126	2.941	3.866	1.154	1.902	2.365	3.250	4.255
45	0.978	1.669	2.092	2.897	3.811	1.122	1.857	2.313	3.181	4.168
50	0.961	1.646	2.065	2.863	3.766	1.096	1.821	2.269*	3.124	4.096

[A]From National Bureau of Standards (refer to bibliography for appendixes).

[B]Factors K such that the probability is γ that at least a proportion P of the distribution will be less than $\bar{X} + Ks$ (or greater than $\bar{X} - Ks$), where $\bar{X}$ and s are estimates of the mean and the standard deviation computed from a sample size of n.

random data collection, unbiased monitoring and analytical procedures, and goodness of fit to assumed parametric distributions should be verified.

Prior to applying inferential tests, the procedures in Appendix II should have been considered and applied where appropriate. Appendix II should be read before Appendixes III and IV. Appendixes II, III, and IV should be read as a unit prior to using the tools in any one appendix.

The goal of the tolerance limit test is to determine with a level of confidence, γ, that at least a proportion, P, of the distribution will be less than a specified exposure level.* The computation is very similar to the confidence interval. K factors are found in Table IV.1 as a function of sample size (n), confidence level (γ), and proportion (P) of the distribution of exposures. For numerical examples in this appendix, the descriptive statistics from Appendix II, with $\gamma = 0.95$, $P = 90\%$, and $n = 17$, should be used.

Tolerance Interval for Normally Distributed Data

When the data are symmetrically distributed and fit a normal distribution (a straight line on linear probability paper), the standard tolerance interval test is appropriate.

If one examines Table IV.1 at $\gamma = 0.95$, $P = 0.90$, and $n = 17$, one will find that $K = 2.001$. On Table II.2, $M = 3.2\ mg/m^3$ and $S = 2.94\ mg/m^3$. The upper tolerance limit (UTL) should be calculated from Equation IV.1.

$$\begin{aligned} UTL &= M + \left(K_{\gamma,P,n}\right)(S) \\ & \quad 3.2 + (2.001)(2.94) \\ &= 9.1\ mg/m^3 \end{aligned} \qquad \textbf{(IV.1)}$$

If one assumes that the population is normally distributed, then the result of applying Equation IV.1 to the 17-sample data

*The tolerance limit is reliable for proportions, P, included within the range between the maximum and minimum data points. Appendix II shows that the usable range for the 17-sample data set is from 5.6% to 94.4%.

set would indicate that one can be 95% confident that 90% of the population of exposures is less than 9.1 mg/m^3. To better understand, one should note that in Figure II.1 the 90th percentile of the sample data is 7.0 mg/m^3. The difference represents the amount of uncertainty about the population distribution that remains after the 17 samples of that population have been evaluated if one assumes normality. If one does not assume normality, the uncertainty is even greater (Appendix VI).

TOLERANCE INTERVAL FOR LOGNORMALLY DISTRIBUTED DATA

When the data are asymmetrically distributed and fit a lognormal distribution (a straight line on log probability paper), a modified tolerance interval test is appropriate. The modification is straightforward; the tolerance interval test is conducted on the logarithms of the data rather than on the data.

If one examines Table IV.1 at $\gamma = 0.95$, $P = 0.90$, and $n = 17$, one will find that $K = 2.001$. From Table II.2, one should recall the mean of the logs, ln(GM) = 0.956, and the standard deviation of the logs, ln(GS) = 0.571. The upper tolerance limit (GUTL) should be calculated from Equation IV.2.

$$\begin{aligned} \ln(\text{GUTL}) &= \ln(\text{GM}) + \left(K_{\gamma,P,n}\right)\ln(\text{GS}) \\ &= 0.956 + (2.001)(0.571) \\ &= 2.01 \qquad \textbf{(IV.2)} \\ \text{GUTL} &= 8.2\ \text{mg/m}^3 \end{aligned}$$

If one assumes that the population is normally distributed, then the result of applying Equation IV.2 to the 17-sample data set would indicate that one can be 95% confident that 90% of the population of exposures is less than 8.2 mg/m^3. In contrast, from Figure II.2, one should note that the 90th percentile of the sample data is 5.4 mg/m^3. The difference represents the amount of uncertainty about the population distribution that remains after the 17 samples of that population have been evaluated if one assumes lognormality. If one does not assume lognormality, the uncertainty is even greater (Appendix VI).

Conclusion

The numerical illustrations were based on descriptive statistics from Table II.2 even though the 17 exposure measurements in those data are not well represented by the parametric assumption. The calculations in Equations III.1 through III.4 can be easily rerun for the refined homogeneous exposure group represented by descriptive statistics from Table II.4.

The reader is encouraged to perform these calculations and to compare results with the summary in Table IV.2. From Table II.4, if one uses n = 15, M = 2.2 mg/m^3, S = 0.417 mg/m^3, ln(GM) = 0.770, ln(GS) = 0.203, one will find $K_{0.95,0.90,15} = 2.068$ from Table IV.1.

This analysis seems to demonstrate 95% confidence that 90% of all exposures in the population represented by all 17 samples is less than about 9.1 mg/m^3. Many industrial hygienists would be inclined to accept this as strong evidence without noticing that the conclusion is violated by the very sample from which it was calculated. Only 15 of 17 samples (88%) are less than 9.1 mg/m^3. The tolerance interval test failed here because, as was demonstrated in Appendix II, the 17-sample data set is not well represented by either normal or lognormal statistics. This illustrates the danger of inferential testing that uses parametric statistics that do not fit the data well. Professional judgment must be applied to each and every analysis—assumptions must be verified.

TABLE IV.2
Tolerance Limit Summary

Assumed Distribution	n[A]	OEL[B] mg/m^3	90%ile mg/m^3	UTL[C] mg/m^3
Normal	17	10	7.0	9.08
Lognormal	17	10	5.4	8.15
Normal	15	10	2.73	3.06
Lognormal	15	10	2.80	3.29

[A]n = Number of samples.
[B]OEL = Occupational exposure limit.
[C]UTL = Upper tolerance limit.

In contrast, the analysis suggests 95% confidence that 90% of all exposures in the population represented by the 15 smallest samples is less than 3.3 mg/m^3. As all exposure measurements in this sample are less than or equal to 2.9 mg/m^3, this conclusion is consistent with the sample from which it was calculated. It, therefore, supports the conclusion in Appendix II that both lognormal and normal statistics represent a good fit to the population of exposures represented by the 15 available samples.

The bottom line is that it is part and parcel of professional judgment to check one's assumptions as one proceeds. Nearly every calculation has internal consistency checks that can help verify one's impressions of the workplace and quantify one's decisions. If the data do not fit a parametric distribution, one may refer to Appendix VI.

Appendix V
Control Chart Analysis

The Objective of Control Chart Analysis

Control chart analysis is a statistical method used to evaluate whether a population of some variable is in "statistical control" (i.e., is not changing over time). As was discussed in the chapter on interpretation and decision making, the assumption of a "stationary" population distribution is a fundamental aspect of performing most statistical tests.

Control charts may be applied to exposure data, concentration data, noise data, or contamination level data to make inferences as to whether the population is in statistical control (i.e., stationary). They can also be used as a semiquantitative visual tool to examine trends over time and to demonstrate magnitudes of these variables relative to past levels or to some standard.

The process is in "statistical control" when repeated samples from the defined population behave as random samples from a stationary population distribution. The tools for using control charts rely on the sampled data exclusively. For control chart analysis, the assumption of randomly collected samples is again central to the analysis.

Information to Be Learned from Control Charts

Control charts provide a running graphical record of small subgroups of randomly collected data from the defined population. Several kinds of control charts can be applied:

1. Average (or mean) chart
2. Range chart (surrogate for standard deviation)

3. Attribute chart (e.g., percent of exposure over guideline; not covered in detail in this book)

Each chart provides unique information. The most commonly applied control chart methods use the mean and range charts on the same data sets. The mean chart gives indications of gross blunders (e.g., outliers), shifts in average exposure, slow trends over time, etc. The range chart provides some information on gross blunders (e.g., outliers), but it is most effective in helping to identify shifts in variability and fast fluctuations (cycles in the process).

Calculation of Control Chart Parameters

The calculation of control chart parameters is rather straightforward. A typical data set suited for control chart analysis has the following characteristics:

k sampling campaigns each of
n random samples from the population

Typically *n* may be from 5 to 10 samples.

From this data set, the calculation of parameters follows these steps:

1. The sample mean and sample range for each of the *k* campaigns should be calculated.

$$\bar{\chi}_k = \sum_{i=1}^{n} \chi_{i,k} \qquad \textbf{(V.1)}$$

$$R_k = \chi_{max,k} - \chi_{min,k} \qquad \textbf{(V.2)}$$

2. The grand mean should be calculated by averaging the means of all *k* campaigns.

$$\bar{\bar{\chi}} = \sum_{i=1}^{k} \bar{\chi}_i \qquad \textbf{(V.3)}$$

3. The sample average range from the *k* campaigns should be calculated.

$$\bar{R} = \sum_{i=1}^{k} R_i \quad \text{(V.4)}$$

4. The mean chart is drawn with the grand mean as centerline and the upper and lower confidence limits are calculated as follows using constants from a control chart table of control chart constants (Table V.1).

$$\text{Centerline} = \bar{\bar{\chi}} \quad \text{(V.5)}$$

$$\text{UCL} = \bar{\bar{\chi}} + A_2\bar{R} \quad \text{(V.6)}$$

$$\text{LCL} = \bar{\bar{\chi}} - A_2\bar{R} \quad \text{(V.7)}$$

$$A_2 = \frac{3}{d_2\sqrt{n}} \quad \text{(V.8)}$$

5. The range chart is constructed using the average range as the centerline and using the following equations and table constants to construct the upper and lower control limits.

$$\text{Centerline} = \bar{R} \quad \text{(V.9)}$$

$$\text{UCL} = D_4\bar{R} \quad \text{(V.10)}$$

$$\text{LCL} = D_3\bar{R} \quad \text{(V.11)}$$

6. Once the graphs have been drawn for the mean and range, the parameter estimates (mean and range) from each of the *k* campaigns should be plotted sequentially (versus time). One should examine the plotted points on the chart to determine if all the points fall within the control limits. If they do not, some investigation of those that do not may be needed. The goal is to

establish the baseline stationary distribution; outliers may need to be omitted from the baseline data set.

7. New data sets of size n should have their means and ranges plotted on the chart sequentially. Do these new points fall within the control limits? If they do, one might decide that the process is in statistical control. If not, the outliers should be investigated. In many cases, values falling outside the control limits indicate that the process may be out of control necessitating action to correct problems (e.g., if exposures are trending higher), or it may indicate that a new baseline needs to be established because the original data are no longer representative of the process.

EXAMPLE

An example of control chart analysis of a given set of wipe testing data is provided here for consideration. These data, given in Table V.2 represent micrograms of agent per 100 cm^2 area wiped (standard procedures followed). There are seven campaigns each consisting of seven samples. The sample mean and range are calculated for each campaign as well as the grand mean and the average range. These summary parameters are combined as follows to find the parameters of the mean chart and range chart. The control part parameters, centerline, UCL, and LCL are plotted for the mean chart and range chart (Figures V.1 and V.2 respectively). Sample mean and range values are plotted sequentially after the control charts are constructed.

$\overline{\chi}$ Chart: Centerline = 29.2 µg
UCL = 29.2 + (0.419) (61.3) = 54.9 µg
LCL = 29.2 – (0.419) (61.3) = 3.5 µg

R Chart: Centerline = 61.3 µg
UCL = 1.924 (61.3) = 117.9 µg
LCL = 0.076 (61.3) = 4.7 µg

A review of these mean and range control charts indicates that the sample taken during the fourth quarter of 1988 was probably an outlier. Housekeeping was found to be poor around that time period, elevating all wipe testing points. Changes were recommended in housekeeping practices as a

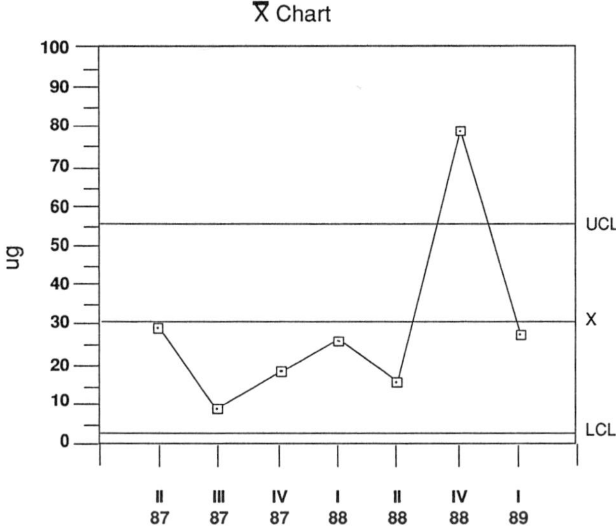

Figure V.1. Mean chart developed as an example of control chart analysis of a given set of wipe testing data

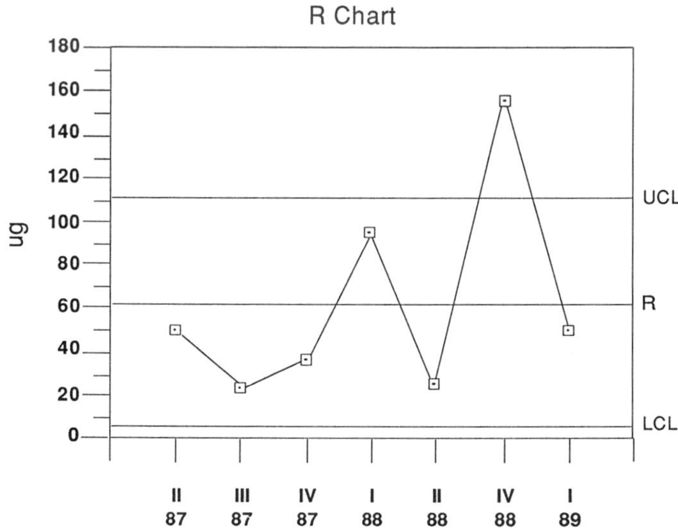

Figure V.2. Range chart developed as an example of control chart analysis of a given set of wipe testing data

TABLE V.1
Control Chart Constants[A]

	$\overline{\chi}$ Chart	R Chart		
	Factors for Control Limits	Factors for Central Line	Factors for Control Limits	
n[B]	A_2[C]	d_2[D]	D_3[E]	D_4[F]
2	1.880	1.128	0	3.267
3	1.023	1.693	0	2.575
4	0.729	2.059	0	2.282
5	0.577	2.326	0	2.115
6	0.483	2.534	0	2.004
7	0.419	2.704	0.076	1.924
8	0.373	2.847	0.136	1.864
9	0.337	2.970	0.184	1.816
10	0.308	3.078	0.223	1.777
11	0.285	3.173	0.256	1.744
12	0.266	3.258	0.284	1.716
13	0.249	3.336	0.308	1.692
14	0.235	3.407	0.329	1.671
15	0.223	3.472	0.348	1.652
16	0.212	3.532	0.364	1.636
17	0.203	3.588	0.379	1.621
18	0.194	3.640	0.392	1.608
19	0.187	3.689	0.404	1.596
20	0.180	3.735	0.414	1.586
21	0.173	3.778	0.425	1.575
22	0.167	3.819	0.434	1.566
23	0.162	3.858	0.443	1.557
24	0.157	3.895	0.452	1.548
25	0.153	3.931	0.459	1.541

[A]Adapted from National Bureau of Standards (refer to bibliography for appendixes).
[B]n = Number of samples in a campaign.
[C]A_2 = Upper confidence limit and lower confidence limit factors for mean chart.
[D]d_2 = Factor for centerline.
[E]D_3 = Lower confidence limit factor for range chart.
[F]D_4 = Upper confidence limit factor for range chart.

control measure. This date's sample should be omitted, and the remaining six campaigns serve as the baseline for the future wipe test results.

TABLE V.2
Example Wipe Test Data for Control Chart Analysis

	Agent on Wipe (μg)						
Quarter Year	II '87	III '87	IV '87	I '88	II '88	IV '88	I '89
Lunchroom	21	7	3.4	8	12.7	30.2	27.5
Engineer's Office	49	9	38	50.5	4.4	112	33.5
Women's Locker Room	0.8	0.21	0.71	2.8	21.5	9.3	8.28
Lab	50	6.6	20	4.4	27.6	119	55.4
Lab Office	29	5	23	3.0	12.7	56	19.9
Control Room	44	23	18	98.9	8.2	77.1	37.3
Men's Locker Room	7.7	6.3	13.4	15.2	20.6	163	18.1
$\overline{\chi}$	28.8	8.2	16.6	26.1	15.4	80.9	28.6
R	49.2	22.8	37.3	95.9	23.2	153.7	47.1

$\overline{\overline{\chi}} = 29.2$ μg
$\overline{R} = 61.3$ μg

CONCLUSION

This appendix has been written to give a flavor of control chart analysis. Correct application will require additional study as well as experience in evaluating real data. Control charting relies on experience and judgment for proper application and interpretation.

Appendix VI
Distribution-Free Tolerance Limit Test

The tolerance limit test applied in Appendix IV failed an internal consistency test because the data from Table II.1 are not well represented by the parametric statistics underlying the test. Nonparametric methods can be used with such data to draw inferences about the workplace. Tables VI.1 and VI.2 allow confidence levels to be estimated for data sets that do not fit parametric assumptions.

NUMBER OF SAMPLES NEEDED

Generally, nonparametric methods require 30 or more samples to achieve reasonable statistical power. Table VI.1 shows that with less than 50 samples, 95% confidence in a one-sided tolerance test is practically unachievable for nonparametric data. In Table VI.2 the number of samples depends both on the proportion P of the population and on the confidence level desired. To achieve 90% confidence that 90% of the population is less than some value requires a random sample of about 40 samples, none of which exceed the value in question.

ONE-SIDED TOLERANCE LIMIT

If one considers the dust exposure data from Appendix II, Table II.1, and assumes that investigation of the two apparent outliers revealed that they are representative of the homogeneous exposure group, then one is left with a data set that does not fit the parametric statistics well enough to calculate confidence levels for tolerance limits. Distribution-free tolerance limit testing is the answer to this dilemma.

TABLE VI.1
Table for Distribution-Free Tolerance Limits (One-Sided)[A,B]

n \ P	γ = 0.75				γ = 0.90				γ = 0.95				γ = 0.99			
	.75	.90	.95	.99	.75	.90	.95	.99	.75	.90	.95	.99	.75	.90	.95	.99
50	10	3	1	—	9	2	1	—	8	2	—	—	6	1	—	—
55	12	4	2	—	10	3	1	—	9	2	—	—	7	1	—	—
60	13	4	2	—	11	3	1	—	10	2	1	—	8	1	—	—
65	14	5	2	—	12	4	1	—	11	3	1	—	9	2	—	—
70	15	5	2	—	13	4	1	—	12	3	1	—	10	2	—	—
75	16	6	2	—	14	4	1	—	13	3	1	—	10	2	—	—
80	17	6	3	—	15	5	2	—	14	4	1	—	11	2	—	—
85	19	7	3	—	16	5	2	—	15	4	1	—	12	3	—	—
90	20	7	3	—	17	5	2	—	16	5	1	—	13	3	1	—
95	21	7	3	—	18	6	2	—	17	5	2	—	14	3	1	—
100	22	8	3	—	20	6	2	—	18	5	2	—	15	4	1	—
110	24	9	4	—	22	7	3	—	20	6	2	—	17	4	1	—
120	27	10	4	—	24	8	3	—	22	7	2	—	19	5	1	—
130	29	11	5	—	26	9	3	—	25	8	3	—	21	6	2	—
140	31	12	5	1	28	10	4	—	27	8	3	—	23	6	2	—
150	34	12	6	1	31	10	4	—	29	9	3	—	26	7	2	—
170	39	14	7	1	35	12	5	—	33	11	4	—	30	9	3	—
200	46	17	8	1	42	15	6	—	40	13	5	—	36	11	4	—
300	70	26	12	2	65	23	10	1	63	22	9	1	58	19	7	—
400	94	36	17	3	89	32	15	2	86	30	13	1	80	27	11	—
500	118	45	22	3	113	41	19	2	109	39	17	2	103	35	14	1
600	143	55	26	4	136	51	23	3	133	48	21	2	126	44	18	1
700	167	65	31	5	160	60	28	4	156	57	26	3	149	52	22	2
800	192	74	36	6	184	69	32	5	180	66	30	4	172	61	26	2
900	216	84	41	7	208	79	37	5	204	75	35	4	195	70	30	3
1000	241	94	45	8	233	88	41	6	228	85	39	5	219	79	35	3

[A]Largest values of m such that one may assert with confidence at least γ that $100P$ percent of a population lies below the mth largest (or above the mth smallest) of a random sample of n from that population (no assumption of normality required).

[B]From National Bureau of Standards (refer to bibliography for appendixes).

Only 17 dust exposure samples were shown in Table II.1. Therefore, the first step would be to collect 33 more samples. With 50 samples in hand, one could enter Table VI.1 at $\gamma = 0.95$, P = 0.90, and n = 50 to find that there is 95% confidence that 90% of the exposures in this population are less than the second largest observed exposure in the data set.

TABLE VI.2
Confidence Associated with a Tolerance Limit Statement[A,B]

n	*P* = .75	*P* = .90	*P* = .95	*P* = .99
3	.16	.03	.01	.00
4	.26	.05	.01	.00
5	.37	.08	.02	.00
6	.47	.11	.03	.00
7	.56	.15	.04	.00
8	.63	.19	.06	.00
9	.70	.23	.07	.00
10	.76	.26	.09	.00
11	.80	.30	.10	.01
12	.84	.34	.12	.01
13	.87	.38	.14	.01
14	.90	.42	.15	.01
15	.92	.45	.17	.01
16	.94	.49	.19	.01
17	.95	.52	.21	.01
18	.96	.55	.23	.01
19	.97	.58	.25	.02
20	.98	.61	.26	.02
25	.99	.73	.36	.03
30	1.00 –	.82	.45	.04
40	—	.92	.60	.06
50	—	.97	.72	.09
60	—	.99	.81	.12
70	—	.99	.87	.16
80	—	1.00 –	.91	.19
90	—	—	.94	.23
100	—	—	.96	.26

[A]Confidence γ with which one may assert that $100P$ percent of the population lies between the largest and smallest of a random sample of n from that population (continuous distribution assumed).

[B]From National Bureau of Standards (refer to bibliography for appendixes).

TWO-SIDED TOLERANCE INTERVAL

If economic constraints preclude additional sampling, then Table VI.2 can be used to make some statements about the underlying population, based on the 17 samples that range from 1.3 to 12 mg/m^3. The row for $n = 17$ indicates that there is 95% confidence that 75% of the population is exposed within this range, 52% confidence that 90% of the population is exposed within this range, 21% confidence that 95% of the population is exposed within this range, and only 1% confidence that 99% of the population is exposed within this range.

Said another way, the data suggest there is a significant proportion of exposures above 12 mg/m^3. Since that is above the occupational exposure limit (OEL), this workplace would be a candidate for additional effort to define the source of unacceptable exposures and implement needed protective measures.

Conclusion

In conclusion, distribution-free statistics are readily available for use. They have much poorer discriminating power than parametric statistics but are a valuable aid to refined professional judgment. When the data do not pass the goodness-of-fit test for parametric statistics, distribution-free statistics are the best alternative.

Beyond that, a closer look at each workplace producing exposure data that fail the goodness-of-fit test is suggested. Often this indicates a workplace with equipment, training, or procedures out of statistical control. Invariably, exposures in such situations can be reduced by enlightened management.

Glossary

absorbed dose The amount of a substance penetrating the exchange boundaries of a worker via either physical or biological processes after contact (exposure).

accuracy The measure of the correctness of data, as given by the difference between the measured value and the true or specified value. Ideal accuracy is zero difference between measured and true value. Contrast with *precision*, the difference between one measured value and the mean of measured values. *See* precision, measurement error.

acute Having a sudden onset, sharp rise, and short course; may apply either to exposures in the workplace or to physiological response.

agent A chemical, radiological, mineralogical, physical, or biological entity that may cause deleterious effects in an exposed worker. Also known as environmental agent or environmental stressor.

analytical methods Chemical analyses; techniques used to quantify an agent collected on sampling media (e.g., gas chromatography/mass spectrometry). Methods used in chemical laboratory analysis of individual hygiene samples.

area sample An environmental sample collected at a fixed point in the workplace; reflects workplace contaminant concentrations that may not correlate with personal samples of individual worker exposure.

arithmetic mean A measure of central tendency, calculated as the sum of all the values in a population divided by the number in the population. Also called mean or average.

asphyxiant A substance that in sufficient concentration renders a person unconscious by preventing sufficient oxygenation of the blood. This may be by chemical or physical blocking of blood oxygenation.

autocorrelation The correlation between successive members of a time series, usually presented as a function of the time separation between members. For a stationary time series, the autocorrelation function is independent of time.

baseline Concentration data that describe the magnitude and range of exposures for a given homogeneous exposure group and agent. There should be at least six samples in a minimum baseline. Descriptive baseline statistics include number of samples, mean, variance, range, and percent of exposures exceeding guideline. The baseline often serves as a comparison for subsequent monitoring data.

bias A systematic error inherent in a method or caused by some feature of the measurement system.

binomial distribution The distribution describing probabilities of outcome of trials each of which can have one of two mutually exclusive results (e.g., measured exposure above or below an occupational exposure limit).

biological monitoring A measurement taken from biological media, usually of the concentration of a chemical/metabolite or the status of a biomarker; normally collected with the intent of relating the measured value to the absorbed dose of a chemical over some averaging period.

biological time constant The time required for a fixed proportion of an absorbed dose to undergo toxicokinetic transformations. Called biological half-life if the proportion equals one-half.

breathing zone A zone of air in the vicinity of a worker from which air is breathed. Personal breathing zone measurements of contaminant concentration are frequently made in occupational health studies by directly placing monitors in the breathing zone of workers.

carcinogen An agent that potentially causes induction of tumors (cancer) following exposure.

ceiling limit Used by the American Conference of Governmental Industrial Hygienists to describe the concentration that should not be exceeded during any part of the working exposure. *See* occupational exposure limit.

central limit theorem The sampling distribution of the mean approaches a normal distribution as the same size increases, regardless of the shape of the underlying population distribution.

certified industrial hygienist An individual who has received the CIH designation from the American Board of Industrial Hygiene (ABIH). To receive ABIH certification, an individual must meet rigorous standards of special education and lengthy experience prior to proving, by written examination, competency in either the comprehensive practice of industrial hygiene or one of the specialties or Aspects (Acoustical, Air Pollution, Chemical, Engineering, Radiological, and Toxicologic). Abbreviated CIH.

chronic Marked by long duration or frequent recurrence; not acute. May refer either to workplace exposures or to the disease or injury state resulting therefrom.

coefficient of variation The sample standard deviation divided by the sample mean (or population parameters). When comparing variation between distributions with different means, coefficients of variation should be used. Sometimes expressed as a percentage. Abbreviated CV.

compliance monitoring Technique for evaluating compliance with government standards; typically, the maximally exposed worker is identified and monitored for exposure to agents. If that personal exposure is less than the standard, then all worker exposures are also assumed to be less.

concentration The amount or volume of a contaminant relative to a given air volume. If measured in the breathing zone of a worker (i.e., a personal sample), it is the number used in comparison with the occupational exposure limit.

confidence interval A range of values (i.e., interval) that has a specified probability of including the true value of a parameter of an underlying distribution. Contrast with the *tolerance interval* that describes the probability that underlying values of a random variable lie within a stated interval. *See* tolerance interval.

confidence level The probability that a stated confidence interval will include a population parameter.

confidence limits The upper and lower boundaries of a confidence interval. Abbreviated upper confidence limit: UCL; lower confidence limit: LCL.

continuous air monitors Direct-reading systems used to monitor air concentrations of agents continuously with some preset averaging time (e.g., consecutive 5-minute time-weighted averages); sometimes linked to alarm systems and trigger alarms if workplace concentrations begin to approach dangerous levels for workers; often used to evaluate concentrations of acute agents (e.g., chlorine) continuously. Abbreviated CAMs.

control chart A chart that shows sequential data or sequential statistics as a function of the time of collection. Usually, decision (control) limits are plotted on the chart to facilitate detection of out-of-control situations (nonstationary distributions).

control strategies Tools to reduce and control workplace agent concentrations and exposures; examples include substitution of a less toxic agent, engineering controls such as ventilation, use of personal protective equipment, administrative controls, and change in work practices.

correlation A number or function with a value between –1 and +1 measuring the relation between variables that tend to vary together in a way not expected on the basis of chance alone. Specifically, the covariance divided by the product of standard deviations.

descriptive statistics Simple ways to characterize a sample distribution's characteristics such as central tendency (e.g., mean, median) and dispersion (e.g., standard deviation, variance, range); includes number of samples, probability plots, and other descriptive tools.

diagnostic monitoring Workplace monitoring conducted to identify sources of unacceptable exposures.

direct measurement of exposure An approach to quantifying exposure by taking measurements of exposure at or near the exchange boundaries of a worker while the exposure is taking place (e.g., the personal air sample in a worker's breathing zone).

dose The amount of a substance available for interaction with metabolic processes of a worker following exposure and absorption. The amount of a substance crossing the exchange boundaries of skin, lungs, or digestive tract is termed *absorbed dose*; the amount available for interaction by any particular organ or cell is termed the *delivered dose* for that organ or cell.

dose rate Dose per unit time (e.g., mg/day) taken up into a worker's body.

dose-response curve A representation, usually in a graphical presentation, of the relationship between dose and probability of occurrence of a health effect or effects. Often these curves are actually based on administered dose (i.e., exposure) rather than absorbed dose or delivered dose.

dosimeter Instrument used to measure dose; many so-called dosimeters actually measure exposure rather than dose.

duration One of three important parameters used to describe the extent and potential consequences of exposures; the other two parameters are frequency and magnitude of exposure.

effective exposure That concentration actually available at the interface of a worker's body (e.g., inside the respirator).

epidemiology The study of the distribution and determinants of health-related states in populations and the application of this study to the control of health problems.

excursion limits Typically, refers to occupational exposure limits applying to exposures of shorter duration than a full-shift time-weighted average or other long-term occupational exposure limit. One type of excursion limit would be a short-term exposure limit. *See* occupational exposure limit.

exposure assessment The determination or estimation (qualitative or quantitative) of the magnitude, frequency, duration, and route of exposure.

exposure assessment strategy A plan to guide industrial hygiene actions and decisions in accomplishing the goal of accurately evaluating the exposures of each worker.

exposure Potential contact at a worker's exchange boundaries (e.g., lung or skin) with chemical, physical, or biological agents. Effective exposure is quantified as the amount of the agent actually available for absorption at the exchange boundaries of the worker's body (e.g., skin, lung tissue, digestive tract, and/or eyes), factoring in proper use of personal protective equipment.

exposure pathway The course a chemical, physical, or biological agent takes from the source to the exposed worker; may be represented by an exposure pathway model.

exposure rate Exposure per unit time (e.g., ppm/hr) potentially at the boundary of a worker's body (e.g., lungs or skin).

exposure route of entry The possible ways a chemical or pollutant enters a worker after contact (e.g., by ingestion, inhalation, or dermal absorption).

exposure scenario A set of assumptions about how exposure takes place (including assumptions/conditions concerning sources, exposure pathways, concentrations of pollutants, individual or population habits and characteristics) that aids the exposure assessor in evaluating, estimating, or quantifying exposures.

extreme value statistic A statistic computed from a sample that estimates a parameter describing a property of the tail(s) of the underlying distribution. Examples of industrial hygiene interest include the 95th percentile (5% exceedance level), the largest values out of a large sample (of interest for substances with ceiling limits), and fraction of exposures that exceed the standard (of interest when decisions are made with the binomial or Poisson distributions.)

environmental agent *See* agent.

fixed-location monitoring Sampling of an environmental or ambient medium for pollution concentration at one location continuously or repeatedly over some length of time.

frequency One of three important parameters used to describe the extent and potential consequences of exposures; the other two parameters are duration and magnitude of exposure.

full shift A shift consisting of regularly scheduled working time (typically, 8 hr).

geometric mean The *n*th root of the product of *n* values. The geometric mean and arithmetic mean of most distributions are not equal and should not be used interchangeably. Arithmetic mean is the correct parameter for evaluating cumulative exposure. The geometric mean is the median of lognormally distributed data.

goodness-of-fit test A formal statistical test that evaluates whether sample data are consistent with the population statistical distribution defined in a null hypothesis.

hazard identification The determination of whether a particular substance or chemical is causally linked to particular health effects (toxicologic assessment).

health-effects data Toxicologic, epidemiologic, and clinical data that demonstrate potential harmful health effects from exposure to environmental agents; often observed in animal studies of agent exposure, although epidemiologic and clinical human data are sometimes available.

homogeneous (Statistics) Having identical probability distribution functions.

homogeneous exposure group A group of employees who experience agent exposures similar enough that monitoring agent exposures of any worker in the group provides data useful for predicting exposures of the remaining workers. Such groups are used in stratified sampling of workplace exposures, thereby improving the power of statistical decision tools. The categorization of workers into such groups often involves categorization by process, job description, and agents, although finer separation can be attained by further dividing on the basis of task analysis. In some cases, workers can be part-time members of different homogeneous exposure groups (e.g., maintenance workers serving several plants). Abbreviated HEG.

hypothesis test A statistical test, based on a random sample from a population, that assigns a confidence level to an assumption or speculation about a parameter of that population (stated in terms of two mutually exclusive but complementary hypotheses). Often in industrial hygiene, a hypothesis test will be performed to compare summary measures of exposure data with standards or occupational exposure limits.

independent Two random variables are said to be independent if the outcome of a trial on one variable does not affect the outcome of a trial on the other and vice versa.

irreversible injury An injury that is not repairable (cannot be expected to heal). Examples include silicosis and permanent hearing loss.

irritant A substance that potentially causes a typically reversible inflammatory reaction in epithelial tissues.

limit of detection The smallest concentration or amount of a substance that can be differentiated from background by a given measurement process (e.g., analytical instrument/ sample combination). Abbreviated LOD.

lognormal distribution The distribution of a random variable that has the property that the logarithms of its values are normally distributed.

long-term average limit A term proposed here to describe the acceptable average concentration of substances with cumulative adverse health effects but no acute effects from short exposures to higher concentrations (no exposure rate dependence). Abbreviated LTA. *See* occupational exposure limit.

magnitude of exposure One of three important parameters used to describe the extent and potential consequences of exposures; the other two parameters are duration and frequency.

maximally exposed individual The single individual with the highest exposure in a given population. Abbreviated MEI.

mean The arithmetic average of a set of data; not used as a substitute for geometric mean.

mean test A statistical test that evaluates whether the sample arithmetic mean exposure equals a specified population mean value.

measurement error The difference between the true value and the value obtained by a measuring device. In the absence of systematic bias, the distribution of measurement errors is often thought of as measurement precision.

median The exposure level that divides a sample into two equal parts—with half less than and half greater than this value.

mode The most frequently occurring value in a set of measurements; the maximum value of a continuous probability density function. The mode of a lognormal distribution is less than the median, which is less than the mean. The mean, median, and mode of a normal distribution are equal. *See* skewed.

model A mathematical physical or subjective representation of real phenomena (e.g., dilution ventilation models).

monitor The act of collecting a sample from the environment for analysis to discover the concentration or intensity of a stressor or environmental agent. Also referred to as monitoring.

monitoring protocols Plans for implementing monitoring campaigns.

nonparametric Statistical methods that do not assume a particular statistical distribution for the statistic of interest (distribution-free methods).

nonrandom sample Any sample taken in such a manner that some members of the defined population may be more likely to be sampled than others.

nonstationary Any process whose statistics vary over time, space, or some other independent variable. Nonstationary processes are not suited to analysis by those statistical tools that explicitly assume an unchanging population distribution.

normal distribution An important symmetric continuous probability distribution characterized completely by two parameters, the mean and the standard deviation. It has its highest ordinate at the center and tails off to zero in both directions, forming a bell-shaped curve. Standard tables of its probability density and probability distribution use the normalized variable, z = (measurement – mean)/standard deviation. Also known as Gaussian distribution. *See* central limit theorem.

null hypothesis The hypothesis about a population parameter to be tested (e.g., the population mean exposure is less than or equal to the occupational exposure limit). The alternative hypothesis covers the complement of the null (e.g., the population mean exposure is greater than the occupational exposure limit). *See* statistical significance.

occupational exposure limit A generic term used to represent a pair of numbers: (1) the agent concentration or intensity that is allowable (based on health-effects data) and (2) the time period over which one averages workplace concentrations to evaluate whether the measured concentrations are less than the allowable limit. Some substances may have several occupational exposure limits (e.g., one for 8 hr, one for 30 min, and a not-to-exceed ceiling). *See* specific types: ceiling limit, permissible exposure limit, threshold limit value, time-weighted average, long-term average, and short-term average. Abbreviated OEL.

overexposure An average exposure greater than the occupational exposure limit. The average should be taken over the appropriate averaging time.

parameter A quantity (such as the mean or the variance) that describes a statistical population. In a computational sense, "nice" distributions are the ones completely characterized by a few parameters. Examples are the normal distribution (characterized by its mean and variance) or lognormal (characterized by its geometric mean and geometric standard deviation).

parametric Described by parameters. Statistical tests are said to be parametric if an underlying statistical distribution, described by appropriate parameters, is based solely on computations involving those parameters.

parametric statistic A statistic that can be computed from estimates of the parameters of the underlying population distribution.

particle-size distribution The statistical distribution of size or mass of aerosols; useful in estimating aerosol exposure to various regions of the respiratory system.

patch samples Use of gauze or other patches to capture aerosolized agents in order to estimate potential deposition of agents onto a worker's skin.

peak exposure The largest exposure or the largest group of exposures experienced by workers during some defined exposure duration (usually short time periods).

permissible exposure limit Defined in 29 CFR 1910, Subpart Z, General Industry Standards for Toxic and Hazardous Substances; usually an 8-hr time-weighted average (TWA) concentration. Abbreviated PEL. *See* occupational exposure limit.

personal measurement A measurement collected from an individual's immediate environment using direct methods such as personal air pumps.

population The collection of all items or elements chosen for study. Usually a small portion of the population is sampled to estimate characteristics of that population. "Nice" populations are those characterized by common probability distributions. The most frequently used in industrial hygiene are the normal and lognormal distributions. Data describing workplace hazards tend to exhibit lognormal-like distributions (skewed); errors involved in monitoring and analyzing exposure samples tend to exhibit normal-like distributions (symmetrical).

population parameters The true parameters calculated by including the entire population (e.g., population mean and standard deviation). In workplace exposure assessment, these parameters must usually be estimated from a representative sample of the population and, therefore, are not known exactly.

potency Health-effect level per unit dose of the agent (in most simple case, the slope of a linear dose-response function).

potential health effects The capability or possibility for altering the health of an individual.

precision A measure of the reproducibility of a measured value under a given set of conditions. Accuracy usually refers to the size of deviations from the true mean, whereas precision refers to the size of deviations from the mean of observations.

predictive exposure assessment An approach to quantifying exposure by measurement or estimation of both the amount of a substance contacted and the frequency/duration of contact, and subsequently linking these together to estimate exposure. Predictive exposure assessments commonly use modeling techniques to make the link between amount (intensity of contact) and duration of contact.

prioritization Placement of industrial hygiene monitoring tasks in rank order of importance (e.g., the rank ordering of homogeneous exposure groups from those needing monitoring immediately down to those groups requiring no monitoring); to arrange in a specified order according to some rank of precedence.

probability density function A function of a continuous random variable whose integral over an interval gives the probability that its value falls within that interval. Also the derivative of the probability distribution function.

probability distribution function A function of a continuous random variable whose value at a point gives the probability that the random variable has a value less than that point. Also the integral of the probability density function.

probability paper Graph paper on which a specific family of distributions is represented by spacing along a probability axis. For example, samples from a lognormal distribution appear as a straight line when plotted on log-probability paper, and samples from a normal distribution appear as a straight line when plotted on linear-probability paper.

probability plot Plot in which sample observations are plotted on the x-axis and the sample probability of occurrence are plotted on the y-axis. If the plot results in a straight line, the sample conforms to the family of distributions represented by the probability paper.

process change Any change in manufacturing, control measures, or procedures that may lead to changes in exposures (exposure distributions).

professional industrial hygienist A person possessing either a baccalaureate degree in engineering, chemistry, or physics or a baccalaureate degree in a closely related biological or physical science from an accredited college or university who also has a minimum of 3 years of industrial hygiene experience. A completed doctorate in a related physical, biological, or medical science or in related engineering can be substituted for 2 years of the 3-year industrial hygiene experience requirement. A completed master's degree in a related physical or biological science or in related engineering can be substituted for 1 year of the 3-year requirement. Under no circumstances can more than 2 years of graduate training be applied toward the 3-year period. While this definition does not include certification, the American Industrial Hygiene Association recognizes the need for such certification by every professional industrial hygienist as an appropriate hallmark by one's peers and strongly urges all eligible members to obtain American Board of Industrial Hygiene certification.

professional judgment That capacity of an experienced professional to draw correct inferences from incomplete quantitative data, frequently on the basis of observations, analogy, and intuition.

qualitative Not based on a rigorous quantitative analysis of data; that is, based on integration of information and judgment.

quality assurance The system of activities whose purpose is to provide to the user of the exposure assessment the assurance that the data used as a basis for a decision meet defined standards of quality (e.g., submittal of spiked or blank samples, lab proficiency testing, and sampling records).

random samples Samples selected from a statistical population such that each sample item has an equal probability of being selected.

range The difference between the largest and smallest values in a measurement data set.

representativeness The degree to which a sample is, or samples are, characteristic of the whole medium, exposure, or dose for which the samples are being used to make inferences.

reversible injury An injury that can be treated so that complete healing takes place.

risk The probability of deleterious health or environmental effects; risk is a function of exposure level and potency (health effect per unit exposure [or dose]).

risk assessment The scientific evaluation (in either qualitative or quantitative terms) of the chance of injury or illness resulting from exposure to a particular form of matter or energy.

sample The portion, part, or subset of the population chosen for statistical analysis and study. Sample data are often used to make inferences about the true values of population parameters. Contrast with *monitor*.

sampler averaging time The "window" or duration over which a sample provides a representative average concentration.

sample parameters Estimators of population parameters based upon observation of a subgroup of the population (e.g., sample mean and sample standard deviation).

sampling frequency The time interval between the collection of successive samples.

short-term exposure limit A term used by the American Conference of Governmental Industrial Hygienists to describe the concentration to which workers can be exposed for a short period of time without exceeding the daily threshold limit value–time-weighted average and without suffering from (1) irritation, (2) chronic or irreversible tissue damage, or (3) narcosis of sufficient degree to increase the likelihood of accidental injury, to impair self rescue, or to reduce work efficiency materially. It is usually defined as a 15-min time-weighted average although other periods are stated when warranted by observed biological effects. Repeat exposures are to be separated by at least 60 min, and no more than 4 such exposures per day are permitted. Abbreviated STEL. *See* occupational exposure limit.

skew A property of a statistical distribution indicating the lack of symmetry; a distribution is said to be skewed to the right or left if observations are concentrated to the left or right of the mean respectively. A distribution is skewed right (positive skew) when mean is greater than the median and is skewed to the left (negative skew) when the mean is less than the median. *See* tail.

skin washes Use of solvent washes (most typically water) to remove agents deposited on a worker's skin; solvent is captured and analyzed for agent presence on skin.

standard deviation The positive square root of the variance of a distribution; the parameter measuring spread of values about the mean; can be estimated from the slope of the straight line through data plotted on probability paper.

standard error The standard deviation of the distribution of a sample statistic. Examples include standard error of the sample mean, standard error of the sample median, standard error of the sample range.

stationary A random process is said to be stationary if its distribution is independent of the time of observation.

statistic A characteristic of a sample such as sample mean and sample standard deviation; used to estimate true values of parameters that characterize the underlying population.

statistical significance Statistical significance is the statistically based probability that the null hypothesis is true. If that probability is small (e.g., less than 5%), the null hypothesis is rejected.

strategy A plan to guide actions to accomplish some stated goal.

stratified sampling A method of sampling in which the population is divided into homogeneous groups, or strata, and elements within each stratum are selected at random.

Student's *t* distribution A family of probability distributions distinguished by their degrees of freedom and used as the sampling distribution of arithmetic means when the population standard deviation is unknown. As sample size increases, the *t* distribution asymptotically approaches a normal distribution.

surveillance Ongoing scrutiny to detect changes in distribution or trends of disease or exposures in order to initiate more focused studies or control measures (e.g., a complete epidemiologic investigation).

systematic sampling A method of selecting samples in which elements to be sampled are selected from the population at a uniform interval measured in terms of an independent variable such as time, order, or space. Large unmeasured error or bias may occur in systematic sample statistics.

tail That part of a distribution represented by extreme values with a low probability of occurrence. *See* skewed.

teratogen An agent whose exposures potentially result in fetal health effects (e.g., birth defects).

threshold limit value Refers to airborne concentrations of substances and represents conditions under which it is believed that nearly all workers may be repeatedly exposed day after day without adverse effect. Abbreviated TLV®. *See* occupational exposure limit.

time-weighted average A term used by the American Conference of Governmental Industrial Hygienists (ACGIH) to describe an average concentration for a normal 8-hr workday and a 40-hr workweek. In applying this guideline, ACGIH suggests that short exposures should not exceed five times the threshold limit value–time-weighted average at any time and should not exceed three times the threshold limit value–time-weighted average more than 30 min per day, and then only if the 8-hr average does not exceed the threshold limit value–time-weighted average. Abbreviated TWA. *See* occupational exposure limit.

tolerance interval An interval that contains a stated fraction of the values of a population distribution with stated probability; refers to the probability that the random variable itself lies within the stated interval. For example, 95% probability that no more than 5% of the daily exposures exceed the standard. Contrast with *confidence interval* that refers to probability (confidence) that a parameter of the underlying distribution lies within the stated interval. *See* confidence interval.

tolerance limit The upper or lower limits of a tolerance interval. Abbreviated upper tolerance limit: UTL; lower tolerance limit: LTL.

toxic Causes deleterious health effects in living organisms.

toxicity The degree to which a chemical or biological agent causes deleterious health effects in living organisms.

toxicodynamics The study of the time course of health effects developing from the delivered doses of the chemical or its metabolites in the organism (effect of chemical and its metabolites on organism).

toxicokinetics The study of the time course of absorption, distribution, metabolism, and excretion of a foreign substance (e.g., a drug or chemical) in an organism's body (effect of organism on chemical).

toxicology The study of the adverse health effects caused by chemicals in humans and animals.

validation Verification of the correctness of the methods used for sampling and laboratory analysis of an agent under conditions of use; also the demonstration that a model accurately predicts real-world phenomena by comparison of model prediction with actual data.

variance The mean of the square of the differences between the mean value of population and randomly selected values from the same population. Units are squares of the data units.

warning properties Properties of an agent that will enable an educated worker to identify potential overexposures (e.g., odor thresholds at or below the occupational exposure limit can alert workers to the agent's presence; however, an odor threshold above the occupational exposure limit means that the worker will have no warning of potential overexposure).

wipe testing The collection of chemical, mineralogical, or radiological agents on wipe media (typically a wipe of filter paper on an area 100 cm^2); results are useful indexes of contamination level but are not direct estimators of exposure.

work history A historical representation of various jobs, process, department, and/or location over an employee's career.

Bibliography

CHAPTER 1: "INTRODUCTION"

Brunn, I.O., J.S. Campbell, and R.L. Hutzel. 1986. Evaluation of Occupational Exposures: A Proposed Sampling Method. *Am. Ind. Hyg. Assoc. J. 47:*229–235.

Chemical Manufacturers Association. 1986. "International Workshop on Strategies for Measuring Exposure," sponsored by the Chemical Manufacturers Association, the American Industrial Hygiene Association, and the American Petroleum Institute, December 9 and 10, 1986, CMA Board Conference Room. CMA, 2501 M Street, NW, Washington, DC 20037.

"Generic Standard for Exposure Monitoring: Advance Notice of Proposed Rulemaking." *Federal Register 52:*(27 Sept. 1988). pp. 37591–37595.

National Institute for Occupational Safety and Health. 1977. *Occupational Exposure Sampling Strategy Manual* by N.A. Leidel, K.A. Busch, and J.R. Lynch (HEW/NIOSH Pub. No. 77-173). Washington, D.C.: Government Printing Office.

Roach, S.A. 1966. A More Rational Basis for Air Sampling Programs. *Am. Ind. Hyg. Assoc. J. 27:*1–12.

Roach, S.A. 1977. A Most Rational Basis for Air Sampling Programmes. *Ann. Occup. Hyg. 20:*65–84.

Roach, S.A. 1987. A Commentary on December 1986 Workshop on Strategies for Measuring Exposure. *Am. Ind. Hyg. Assoc. J. 48(12):* A-822–A-832.

Rock, J.C. 1982. A Comparison between OSHA-Compliance Criteria and Action-Level Decision Criteria. *Am. Ind. Hyg. Assoc. J. 43:*297–313.

Rock, J.C. 1986. Can Professional Judgment Be Quantified? *Am. Ind. Hyg. Assoc. J. 47:*A-370.

Schulte, H.F. 1962. Modern Concepts of Air Sampling and Problems for the Future. *Am. Ind. Hyg. Assoc. J. 23:*20–25.

Zielhuis, R.L., P.C. Noordam, H. Roelfzema, and A.A.E. Wibowo. 1988. Short-Term Occupational Exposure Limits—A Simplified Approach. *Int. Arch. Occup. Environ. Health 61:*207–211.

CHAPTER 2: "BASIC CHARACTERIZATION"

American Conference of Governmental Industrial Hygienists. 1986. *Documentation of the Threshold Limit Values for Substances in Workroom Air*. Fifth ed. Cincinnati, Ohio: American Conference of Governmental Industrial Hygienists.

American Conference of Governmental Industrial Hygienists. 1986. *Industrial Ventilation, A Manual of Recommended Practice*. Nineteenth ed. Cincinnati, Ohio: American Conference of Governmental Industrial Hygienists.

American Industrial Hygiene Association. 1977. Nonionizing Radiation Guide Series. Akron, Ohio: American Industrial Hygiene Association.

American Industrial Hygiene Association. 1986. *Noise and Hearing Conservation Manual* edited by E.H. Berger, W.D. Ward, J.C. Morrill, and L.H. Royster. Fourth ed. Akron, Ohio: American Industrial Hygiene Association.

Checkoway, H., J.M. Dement, D.P. Fowler, R.L. Harris, S.H. Lamm, and T.J. Smith. 1987. Industrial Hygiene Involvement in Occupational Epidemiology. *Am. Ind. Hyg. Assoc. J. 48(6)*:515–523.

Chemical Manufacturers Association. 1986. "International Workshop on Strategies for Measuring Exposure," sponsored by the Chemical Manufactures Association, the American Industrial Hygiene Association, and the American Petroleum Institute, December 9 and 10, 1986, CMA Board Conference Room. CMA, 2501 M Street, NW, Washington, DC 20037.

Corn, M. and N. Esmen. 1979. Workplace Exposure Zones for Classification of Employee Exposures to Physical and Chemical Agents. *Am. Ind. Hyg. Assoc. J. 40*:47–54 (1979).

Cralley, J.V. and Cralley, L.J. (eds.). 1985. *Theory and Rationale of Industrial Hygiene Practice, Patty's Industrial Hygiene and Toxicology*. R.L. Harris, Jr., and E.W. Arp, Jr. *The Emission Inventory, Vol. IIIA*, Second ed. New York: John Wiley and Sons. pp. 537–567.

Cralley, L.V. and Cralley, L.J. (eds.). 1987. *Industrial Hygiene Aspects of Plant Operations*. New York: Macmillan.

Dialog Information Services, Inc. 1988. Occupational Safety and Health (NIOSH) Information Retrieval Service, Dialog File #161. Palo Alto, California. June 1988.

Doull, J., C.D. Klaassen, and M.O. Amdur (eds). 1980. *Casarett and Doull's Toxicology*. New York: Macmillan Publishing Co., Inc.

Hansen, D.J. 1986. "Exposure Group Determination in the Solvent Spraying Industries Using Simple Survey Methods." Ph.D. diss., The University of Michigan, Ann Arbor, Michigan.

Hansen, D.J., J.A. Foulke, and R.A. Deininger. 1985. The Portable Computer as a Time Study Tool. In *Softcover Software*, edited by G.E. Whitehouse. Atlanta: Industrial Engineering Management Press. pp. 1–6.

Hansen, D.J. and L.W. Whitehead. 1988. The Influence of Task and Location on Solvent Exposures in a Printing Plant. *Am. Ind. Hyg. Assoc. J. 49(5):* 259–265.

Hansen, D.J., G.W. Adams, and R.A. Hochberg. 1989. Evaluation of Sampling Strategies Used during a Noise Survey of a Machine Shop. *Appl. Ind. Hyg. 4(3):*75–80.

National Library of Medicine. Hazardous Substances Data Bank (HSDB). Operates off the TOXNET System, Bethesda, Md. (Contains a variety of reference sources.)

Langer, R.R., S.K. Norwood, G.E. Socha, and H.R. Hoyle. 1979. Two Methods of Establishing Industrial Hygiene Priorities. *Am. Ind. Hyg. Assoc. J. 40:*1039–1045.

Lundberg, I., B. Sjogren, U. Hallane, L. Hedstrom, and M. Holgersson. 1984. Environmental Factors and Uptake of Cadmium among Brazers Using Cadmium-Containing Hard Solders. *Am. Ind. Hyg. Assoc. J. 45:*353–359.

Nigg, H.N., J.H. Stamper, and R.M. Queen. 1984. The Development and Use of a Universal Model to Predict Tree Crop Harvester Pesticide Exposure. *Am. Ind. Hyg. Assoc. J. 45:*182–186.

Parmeggiani, L. (ed.). 1983. *Encyclopedia of Occupational Health and Safety.* Third rev. ed. Geneva, Switzerland: International Labour Office.

Perry, R.H., D.W. Green, and J.O. Maloney (eds.). 1984. *Chemical Engineers' Handbook.* Sixth ed. New York: McGraw Hill.

Sax, N.I. and Lewis, R.J. 1987. *Hawley's Condensed Chemical Dictionary.* Eleventh ed. New York: Van Nostrand Reinhold.

Todd, W.F., and S.A. Shulman. 1984. Control of Styrene Vapor in a Large Fiberglass Boat Manufacturing Operation. *Am. Ind. Hyg. Assoc. J. 45:* 817–825.

Tucker, M.E. 1984. *Industrial Hygiene: A Guide to Technical Information Sources.* Akron, Ohio: American Industrial Hygiene Association.

Weast, R. (ed.). 1987–1988. *Handbook of Chemistry and Physics.* Sixty-eighth ed. Boca Raton, Florida: CRC Press, Inc.

Chapter 3: "Qualitative Risk Assessment and Exposure Ranking"

American Conference of Governmental Industrial Hygienists. 1986. *Industrial Ventilation: A Manual of Recommended Practice.* 19th ed. Cincinnati, Ohio: American Conference of Governmental Industrial Hygienists.

Chemical Manufacturers Association. 1986. "International Workshop on Strategies for Measuring Exposure," sponsored by the Chemical Manufacturers Association, the American Industrial Hygiene Association, and the American Petroleum Institute, December 9 and 10, 1986, CMA Board Conference Room. CMA, 2501 M Street, NW, Washington, DC 20037.

Corn, M. and N. Esmen. 1979. Workplace Exposure Zones for Classification of Employee Exposures to Physical and Chemical Agents. *Am. Ind. Hyg. Assoc. J. 40*:47–54.

Damiano, J. 1989. A Guideline for Managing the Industrial Hygiene Sampling Function. *Am. Ind. Hyg. Assoc. J. 50(7)*:366–371.

Gamble, J. and R. Spirtas. 1976. Job Classification and Utilization of Complete Work Histories in Occupational Epidemiology. *J. Occup. Med. 18*:399–404.

Gibbs, G.W. and M. LaChance. 1972. Dust Exposure in the Chrysotile Asbestos Mines and Mills of Quebec. *Arch. Environ. Health 24*:189–197.

Hansen, D.J. 1986. "Exposure Group Determination in the Solvent Spraying Industries Using Simple Survey Methods." Ph.D. diss., The University of Michigan, Ann Arbor, Michigan.

Hansen, D.J. and L.W. Whitehead. 1988. The Influence of Task and Location on Solvent Exposures in a Printing Plant. *Am. Ind. Hyg. Assoc. J. 49(5)*: 259–265.

Hawkins, N.C. and J.S. Evans. 1989. Subjective Estimation of Toluene Exposures: A Calibration Study of Industrial Hygienists. *Appl. Ind. Hyg. 4(3)*:61–68.

Jodoin, G., G.W. Gibbs, P.T. Macklem, J.C. McDonald, and M.R. Becklake. 1971. Early Effects of Asbestos Exposure on Lung Function. *Am. Rev. Respir. Dis. 104*:525–535.

Kromhout, H., Y. Oostendorp, D. Heederik, and J.S.M. Boleij. 1987. Agreement between Qualitative Exposure Estimates and Quantitative Exposure Measurements. *Am. J. Ind. Med. 12*:551.

Langner, R.R., S.K. Norwood, G.E. Socha, and H. Hoyle. 1979. Two Methods for Establishing Industrial Hygiene Priorities. *Am. Ind. Hyg. Assoc. J. 40*:1039–1045.

Larson, M., R. Wolford, and T. Crancer. 1982. "IBPAT Solvent Hazard Index." Paper presented at the American Industrial Hygiene Conference, Cincinnati, Ohio, June 6–11, 1982.

Lemasters, G.K., A. Carson, and S.J. Samuels. 1985. Occupational Exposure for Twelve Product Categories in the Reinforced-Plastics Industry. *Am. Ind. Hyg. Assoc. J. 46*:434–441.

Mansdorf, S.Z., A.L. Lott, and T.W. Knupp. 1987. "Development of a Risk Management Database." Paper presented at the American Industrial Hygiene Conference, Montreal, Canada, May 31–June 5, 1987.

McDonald, J.C., A.D. McDonald, G.W. Gibbs, J. Siemiatycki, and C.E. Rossiter. 1971. Mortality in the Chrysotile Asbestos Mines and Mills of Quebec. *Arch. Environ. Health 22*:677–686.

McMichael, A.J., R. Spirtas, L.L. Kupper, and J.F. Gamble. 1975. Solvent Exposure and Leukemia among Rubber Workers: An Epidemiological Study. *J. Occup. Med. 17*:234–239.

Murphy, D.C. 1984. Acute Illness among Workers Connected to Solvent Exposure. *Occup. Health Saf. 53*:36–38.

Ott, W.R. 1980. "Models of Human Exposure to Air Pollution," (Technical Report No. 32). SIAM Institute for Mathematics and Society, Dept. of Statistics, Stanford University, Palo Alto, CA 94305.

Ouellette, R.P., A.I. Garte, and C.W. Thompson. 1985. Assessing Asbestos Risks Involves Prioritizing, Ranking Hazard Areas. *Occup. Health Saf. 55*:67–72.

Van Ert, M.D., E.A. Arp, R.L. Harris, M.J. Symons, and T.M. Williams. 1980. Worker Exposures to Chemical Agents in the Manufacture of Rubber Tires: Solvent Vapor Studies. *Am. Ind. Hyg. Assoc. J. 41*:212–219.

Wadden, R.A. and P.A. Scheff. 1983. *Indoor Air Pollution—Characterization Prediction, and Control.* New York: John Wiley and Sons.

Winn, D., W. Lednar, T. Williams, and M. Van Ert. 1977. "The Identification of Occupational Exposure by an Industrial Hygiene Questionnaire." Occupational Health Studies Group, University of North Carolina School of Public Health. Presented to the American Industrial Hygiene Conference, New Orleans, Louisiana, May 22–27, 1977.

Zarouri, M.D., R.J. Heinsohn, and C.L. Merkle. 1983. Computer-Aided Design of a Grinding Booth for Large Castings. *ASHRAE Trans. 2A*:95–118.

Zarouri, M.D., R.J. Heinsohn, and C.L. Merkle. 1983. Numerical Computation of Trajectories and Concentrations of Particles in a Grinding Booth. *ASHRAE Trans. 2A*:119–135.

CHAPTER 4: "MONITORING"

Aitto, A., J. Jarvisalo, V. Riihimaki, and S. Hernberg. 1988. Biological Monitoring. In *Occupational Medicine Principles and Practical Applications*, edited by C. Zenc. Second ed. Chicago: Yearbook Medical Publishers.

American Industrial Hygiene Association. Nonionizing Radiation Guide Series. Akron, Ohio: American Industrial Hygiene Association.

American Industrial Hygiene Association. 1986. *Noise and Hearing Conservation Manual*, edited by E.H. Berger, W.D. Ward, J.C. Morrill, and L.H. Royster. Fourth ed. Akron, Ohio: American Industrial Hygiene Association.

American Industrial Hygiene Association. 1988. *Quality Assurance Manual for Industrial Hygiene Chemistry*. Akron, Ohio: American Industrial Hygiene Association.

Atherley, G. 1985. A Critical Review of Time-Weighted Average as an Index of Exposure and Dose and of Its Key Elements. *Am. Ind. Hyg. Assoc. J. 46*:481–487.

Baselt, R.C. 1988. *Biological Monitoring Methods for Industrial Chemicals*, 2d ed. Littleton, Mass.: PSG Publishing Company.

Behar, A. and R. Plenar. 1984. Sampling Strategy and Risk Assessment. *Am. Indus. Hyg. Assoc. J. 45:*105–109.

Breslin, A.J.: Solving Air Contamination Problems through Diagnostic Air Sampling. *Am. Ind. Hyg. Assoc. J. 27:*460–468.

Chemical Manufacturers Association. 1986. "International Workshop on Strategies for Measuring Exposure," sponsored by the Chemical Manufacturers Association, the American Industrial Hygiene Association, and the American Petroleum Institute, December 9 and 10, 1986, CMA Board Conference Room. CMA, 2501 M Street, NW, Washington, DC 20037.

Corn, M. 1984. Air Sampling Strategies in the Work Environment. *Am. J. Ind. Med. 6:*251–252. [Editorial].

Cralley, L.J. and L.V. Cralley. (eds.). 1985. *Theory and Rationale of Industrial Hygiene Practice, Patty's Industrial Hygiene and Toxicology.* N.A. Leidel and K.A. Busch. *Statistical Design and Data Analysis Requirements. Volume IIIA,* New York: John Wiley and Sons. pp. 395–508.

Department of Health, Education, and Welfare (Public Health Service). 1973. *The Industrial Environment—Its Evaluation and Control* (CDC, NIOSH). Washington, D.C.: Government Printing Office.

Gerber, W.S., E.W. Arp, R.J. Kopec, and L.W. Carstensen. 1978. "Computer Mapping of Industrial Hygiene Sampling Data." Paper presented at the American Industrial Hygiene Conference, Los Angeles, California, May 11, 1978.

Gifford, F.A., Jr. 1961. Use of Routine Meteorological Observations for Estimating Atmospheric Dispersion. *Nuc. Saf. 2:*47–51.

Gilbert, R.O. 1987. *Statistical Methods for Environmental Pollution Monitoring.* New York: Van Nostrand Reinhold Co.

Hansen, D.J. 1986. "Exposure Group Determination in the Solvent Spraying Industries Using Simple Survey Methods." Ph.D. diss., The University of Michigan, Ann Arbor, Michigan.

Hansen, D.J. and L.W. Whitehead. 1988. The Influence of Task and Location on Solvent Exposures in a Printing Plant. *Am. Ind. Hyg. Assoc. J. 49(5):* 259–265.

Hansen, D.J., G.W. Adams, and R.A. Hockberg. 1989. Evaluation of Sampling Strategies Used during a Noise Survey of a Machine Shop. *Appl. Ind. Hyg. 4(3):*75–80.

Hansen, D.J., R.G. Radwin, T.J. Armstrong, and R.P. Garrison. 1990. Computer-Aided Design (CAD) in Industrial Hygiene. In *Computers in Health and Safety,* ed. by G.M. Rawls. Akron, Ohio: American Industrial Hygiene Association.

Heggle, A.S. 1983. Employee Movement Analysis. *Occup. Health Saf. 52:*31–36.

Hinds, W.C. 1982. *Aerosol Technology.* New York: John Wiley and Sons.

Jones, B. and R.L. Harris. 1984. Calculation of Time-Weighted Average Concentrations: A Computer Mapping Application. *Am. Ind. Hyg. Assoc. J. 44:*795–801.

Klassen, C.D., M.O. Amdur, and J.D. Doull. 1986. *Cassarett and Doull's Toxicology*. Third ed. New York: Macmillan.

Lundberg, I., B. Sjogren, U. Hallane, L. Hedstrom, and M. Holgersson. 1984. Environmental Factors and Uptake of Cadmium among Brazers Using Cadmium-Containing Hard Solders. *Am. Ind. Hyg. Assoc. J. 45:*353–359.

National Institute for Occupational Safety and Health. 1977. *Occupational Exposure Sampling Strategy Manual*, by N.A. Leidel, K.A. Busch, and J.R. Lynch (HEW/NIOSH Pub. No. 77-173). Washington, D.C.: Government Printing Office.

Nigg, H.N., J.H. Stamper, and R.M. Queen. 1984. The Development and Use of a Universal Model to Predict Tree Crop Harvester Pesticide Exposure. *Am. Ind. Hyg. Assoc. J. 45:*182–186.

Pasquill, F. 1961. The Estimation of the Dispersion of Windborne Material. *Meteorol. Mag. 90:*33–49.

Pasquill, F. 1962. *Atmospheric Diffusion*. London: Van Nostrand.

Popendorf, W. 1984. Vapor Pressure and Solvent Vapor Hazards. *Am. Ind. Hyg. Assoc. J. 45:*719–726.

Powell, R.W. 1984. Estimating Worker Exposure to Gases and Vapors Leaking from Pumps and Valves. *Am. Ind. Hyg. Assoc. J. 45:*A-7–A-15.

Rappaport, S.M. 1987. Smoothing of Exposure Variability at the Receptor: Implications for Health Standards. *Ann. Occup. Hyg. 29:*201–214.

Rappaport, S.M., R.C. Spear, and J.W. Yager. 1982. Industrial Hygiene Data: Compliance, Dosage, and Clinical Relevance. *West. J. Med. 132:*572–576.

Rock, J.C. 1981. "An Industrial Hygienist's View: Relationships Between Industrial Hygiene Sampling, Human Exposure Standards, and Experimental Toxicology." Paper #11, proceedings of the Twelfth Annual Conference on Environmental Toxicology. Dayton, Ohio, November 1981.

Saltzman, B.E. 1985. Variability and Bias in the Analysis of Industrial Hygiene Samples. *Am. Ind. Hyg. Assoc. J. 46:*134–141.

Saltzman, B.E. 1988. Linear Pharmacokinetic Models for Evaluating Unusual Work Schedules, Exposure Limits, and Body Burdens of Pollutants. *Am. Ind. Hyg. Assoc. J. 49:*213–225.

Schroy, J.M. 1981. Prediction of Workplace Contaminant Levels. In *Symposium Proceedings: Control Technology in the Plastics and Resins Industry* (DHHS/NIOSH Pub. No. 80-107). Washington, D.C.: Government Printing Office.

Spear, R.C., S. Selvin, and M. Francis. 1986. The Influence of Averaging Time on the Distribution of Exposures. *Am. Ind. Hyg. Assoc. J. 47(6):*365–368.

Ulfvarson, U. 1984. Limitations to the Use of Employee Exposure Data on Air Contaminants in Epidemiologic Studies. *Int. Arch. Occup. Environ. Health 52:*285–300.

Ulfvarson, U. 1987. Assessment of Concentration Peaks in Setting Exposure Limits for Air Contaminants at Workplaces, with Special Emphasis on Narcotic and Irritative Gases and Vapors. *Scand. J. Work, Environ. Health 13*:389–398.

Welling, G.P. 1986. *Pharmacokinetics, Processes, and Mathematics*. ACS Monograph 185. Washington, D.C.: American Chemical Society.

Williams, P.L., M.T. Luster, and P.J. Middendorf. 1983. Area Sampling Methods: Selecting the Proper Sites. *Occup. Health Saf. 53*:21–24.

CHAPTER 5: "INTERPRETATION AND DECISION MAKING"

Aitchison, J. and J.A.C. Brown. 1957. *The Lognormal Distribution—With Special Reference to Its Use in Economics*. Cambridge, United Kingdom: Cambridge University Press.

Amstadter, B.L. 1971. *Reliability Mathematics: Fundamentals; Practices; Procedures*. New York: McGraw-Hill Book Co.

Bar-Shalom, Y., A. Segall, and D. Budenaers. 1976. Decision and Estimation Procedures for Air Contaminants. *Am. Ind. Hyg. Assoc. J. 37*:469–473.

Box, G.E.P., W.G. Hunter, and J.S. Hunter. 1978. *Statistics for Experimenters*. New York: John Wiley and Sons, Inc.

Burlington, R.S. and D.C. May. 1970. *Handbook of Probability and Statistics with Tables*, 2d ed. New York: McGraw-Hill Book Co.

Chambers, J.M., W.S. Cleveland, B. Kleiner, and P.A. Tukey. 1985. *Graphical Methods for Data Analysis*. Belmont, Calif.: Wadsworth International Group and Duxbury Press.

Chemical Manufacturers Association. 1986. "International Workshop on Strategies for Measuring Exposure," sponsored by the Chemical Manufacturers Association, the American Industrial Hygiene Association, and the American Petroleum Institute, December 9 and 10, 1986, CMA Board Conference Room. CMA, 2501 M Street, NW, Washington, D.C. 20037.

Coenen, W. 1966. The Confidence Limits for the Mean Values of Dust Concentrations. *Staub. 37(3)*:39–45.

Conover, W.J. 1980. *Practical Nonparametric Statistics*, Second ed. New York: John Wiley and Sons.

Environmental Protection Agency. 1980. *Health Physics Society Committee Report, Upgrading Environmental Radiation Data* (HPSR-1) (USEPA 520/1-80-012). Washington, D.C.: Government Printing Office.

Esmen, N. and Y. Hammad. 1977. Lognormality of Environmental Sampling Data. *Environ. Sci. Health A12*:29–41.

Evans, J.S. and N.C. Hawkins. 1988. The Distribution of Student's t-Statistic for Small Samples from Lognormal Exposure Distributions. *Am. Ind. Hyg. Assoc. J. 49*:512–515.

Filliben, J.J. 1975. The Probability Plot Correlation Coefficient Test for Normality. *Technometrics 17*:111.

Gilbert, R.O. 1987. *Statistical Methods for Environmental Pollution Monitoring.* New York: Van Nostrand Reinhold Co.

Gumbel, E.J. 1958. *Statistics of Extremes.* New York: Columbia University Press.

Herbach, L. 1984. Introduction, Gumbel Model. In *Statistical Extremes & Applications,* Proceedings of the NATO Advanced Study Institute on Statistical Extremes and Applications, Vimerio, Portugal, 31 Aug–14 Sept., 1983, edited by J. Tiago de Oliveira. Dordrecht, Holland: D. Reidel Publishing Company. pp. 49–80.

Jones, A.R. and R.S. Brief. 1971. Evaluating Benzene Exposures. *Am. Ind. Hyg. Assoc. J. 32*:610–613.

Kerr, G.W. 1962. Use of Statistical Methodology in Environmental Monitoring. *Am. Ind. Hyg. Assoc. J. 23*:75–82.

Larsen, R.I. 1969. A New Mathematical Model of Air Pollutant Concentration Averaging Time and Frequency. *J. Air Pollut. Control. Assoc. 19(1)*:24–30.

Leidel, N.A. and K.A. Busch. 1985. Statistical Design and Data Analysis Requirements. In *Patty's Industrial Hygiene and Toxicology, Volume 3A,* 2d ed., edited by L.J. Cralley and L.V. Cralley. New York: John Wiley and Sons, Inc. Chapter 8.

Levin, R.I. 1984. *Statistics for Management,* 3d ed. Englewood Cliffs, N.J.: Prentice-Hall, Inc.

Mostellor, F. and R.E.K. Roarke. 1973. *Sturdy Statistics: Nonparametrics and Order Statistics.* Reading, Mass.: Addison-Wesley Publishing Company.

National Bureau of Standards. 1963. *Experimental Statistics. National Bureau of Standards Handbook 91,* by M.G. Natrella. Washington, D.C.: National Bureau of Standards.

National Institute for Occupational Safety and Health. 1975. *Handbook of Statistical Tests for Evaluating Employee Exposure to Air Contaminants,* by Y. Bar-Shalom, D. Budenaers, R. Schainker, and A. Segall (HEW/NIOSH Pub. No. 75-147). Washington, D.C.: Government Printing Office.

National Institute for Occupational Safety and Health. 1975. *Statistical Methods for the Determination of Noncompliance with Occupational Health Standards,* by N.A. Leidel and K.A. Busch (HEW/NIOSH Pub. No. 75-159). Washington, D.C.: Government Printing Office.

National Institute for Occupational Safety and Health. 1980. *Statistical Properties of Air Sampling Strategies,* by D.H. Budenaers, R.W.J. Lau, and R.N. Patton (NIOSH Pub., NTIS PB83-135954). Washington, D.C.: Government Printing Office.

Norwood, S.K. 1984. "Estimating Employee Exposures from Continuous Air Monitoring Data." Ph.D. diss., University of Oklahoma, Norman, Oklahoma.

"Occupational Exposure to Benzene: Final Rule." *Federal Register 52:*(11 Sept. 1987). pp. 34460–34578.

Oldham, P.D.: The Nature of the Variability of Dust Concentrations at the Coal Face. *Br. J. Ind. Med. 10:*227–234.

Rappaport, S.M. and S. Selvin. 1987. A Method for Evaluating the Mean Exposure from a Lognormal Distribution. *Am. Ind. Hyg. Assoc. J. 48:* 374–379.

Rappaport, S.M., S. Selvin, and S.A. Roach. 1988. A Strategy for Assessing Exposures with Reference to Multiple Limits. *Appl. Ind. Hyg. 3(11):*310–315.

Roach, S.A., E.J. Baier, H.E. Ayer, and R.L. Harris. 1967. Testing for Compliance with Threshold Limit Values for Respirable Dusts. *Am. Ind. Hyg. Assoc. J. 28:*543–553.

Rushton, S. 1950. On a Sequential t-Test. *Biometrika 37:*326.

Saltzman, B.E. 1987. Lognormal Model for Health Risk Assessment of Fluctuating Concentrations. *Am. Ind. Hyg. Assoc. J. 48:*140–149.

Selvin, S., S.M. Rappaport, R. Spear, J. Schulman, and M. Francis. 1987. A Note on the Assessment of Exposure Using One-Sided Tolerance Limits. *Am. Ind. Hyg. Assoc. J. 48:*89–93.

Travis, C.T. and M.L. Land. 1990. Estimating the Mean of Data Sets with Nondetectable Values. *Environ. Sci. Technol. 24(7):*61–63.

Tuggle, R.M. 1981. The NIOSH Decision Scheme. *Am. Ind. Hyg. Assoc. J. 42:*493–498.

Tuggle, R.M. 1982. Assessment of Occupational Exposure Using One-Sided Tolerance Limits. *Am. Ind. Hyg. Assoc. J. 43:*338–346.

Ulfvarson, U. 1977. Statistical Evaluation of the Results of Measurements of Occupational Exposure to Air Contaminants. *Scand. J. Work, Environ. Health 3:*109–115.

Wald, A. 1947. *Sequential Analysis.* New York: John Wiley & Sons.

Zar, J. 1984. *Biostatistical Applications.* 2d ed. New Jersey: Prentice-Hall.

CHAPTER 6: "RECOMMENDATIONS AND REPORTING"

American Chemical Society. 1988. *Basics of Technical Communicating*, edited by B.E. Cain. Washington, D.C.: American Chemical Society.

Chemical Manufacturers Association. 1986. "International Workshop on Strategies for Measuring Exposure," sponsored by the Chemical Manufacturers Association, the American Industrial Hygiene Association, and the American Petroleum Institute, December 9 and 10, 1986, CMA Board Conference Room. CMA, 2501 M Street, NW, Washington, D.C. 20037.

Garrett, J.T., L.J. Cralley, and L.V. Cralley (eds.). 1988. *Industrial Hygiene Management*. New York: John Wiley and Sons.

National Safety Council. 1979. *Fundamentals of Industrial Hygiene*, edited by J. Olishifski. Second ed. Chicago, Ill.: National Safety Council.

Ott, M.G., S.K. Norwood, and R.R. Cook. 1985. The Collection and Management of Occupational Exposure Data. *Am. Stat. 39*:432–436.

Rathbone, R.R. 1966. *Communicating Technical Information: A Guide to Current Uses and Abuses in Scientific and Engineering Writing*. Reading, Mass.: Addison-Wesley.

Wrench, C. 1989. *Data Management for Occupational Health and Safety: A User's Guide to Integrating Software*. New York: Van Nostrand Reinhold.

CHAPTER 7: "REEVALUATION"

Chemical Manufacturers Association. 1986. "International Workshop on Strategies for Measuring Exposure," sponsored by the Chemical Manufacturers Association, the American Industrial Hygiene Association, and the American Petroleum Institute, December 9 and 10, 1986, CMA Board Conference Room. CMA, 2501 M Street, NW, Washington, D.C. 20037.

Damiano, J. 1989. A Guideline for Managing the Industrial Hygiene Sampling Function. *Am. Ind. Hyg. Assoc. J. 50(7)*:366–371.

APPENDIXES

Aitchison, J. and J.A.C. Brown. 1976. *The Lognormal Distribution*. London: Cambridge University Press.

Box, G.E.P., W.G. Hunter, and J.S. Hunter. 1978. *Statistics for Experimenters*. New York: John Wiley and Sons.

Cochran, W.G. 1953. *Sampling Techniques*. New York: John Wiley and Sons, Inc.

Gumbel, E.J. 1958. *Statistics of Extremes*. New York: Columbia University Press.

Hawkins, N.C. and B.D. Landenberger. 1991. Statistical Control Charts: A Technique for Analyzing Industrial Hygiene Data. *Appl. Occup. Environ. Hyg. 6 (8)*: 689–695.

Hines, W.W. and D.C. Montgomery. 1980. *Probability and Statistics in Engineering and Management Science. Second Edition*. New York: John Wiley and Sons.

National Bureau of Standards. 1966. *Experimental Statistics. National Bureau of Standards Handbook 91*. Washington, D.C.: National Bureau of Standards.